Carbonomics

According to many of my friends, the cover should have pictured a polar bear stranded on a tiny, shrinking iceberg. True, climate change will harm polar bears, and perhaps they will survive global warming only in zoos. But this book focuses on the power of the polar bear.

Ironically, the Organization of Petroleum Exporting Countries (OPEC) understands this power better than many environmentalists do. In fact, OPEC is fighting the bear—currently represented rather poorly by the Kyoto Protocol—tooth and nail. I am not talking about OPEC's high prices, which actually help the bear, but rather its opposition to carbon taxes.

For over three decades, concerns about energy security have failed to stimulate the global cooperation required to defeat OPEC. But concerns about climate are having a greater impact. If environmentalists and proponents of energy security realize that the polar bear is their one best hope, they can save the climate, defeat OPEC, and save consumers hundreds of billions of dollars a year in payments to OPEC and Big Oil. This book explains how.

Below the bear, a few OPEC ministers pose for a photo during the 146th OPEC conference in Abu Dhabi, United Arab Emirates, on December 5, 2007. The *New York Times* reported that "Saudi Arabia had initially said the group might consider increasing output by 500,000 barrels a day, but backed down amid the opposition of other OPEC members." Apparently OPEC was satisfied with the price at that time—$88 a barrel. No doubt, they were delighted when the price passed $100 a barrel a short time later.

OPEC photo by Kamran Jebreili/AP Images. The polar bear photo is owned by the Aflo Company, which provides no information about the bear, although it is almost certainly in a zoo.

Carbonomics
How to Fix the Climate and Charge It to OPEC

by

Steven Stoft

Carbonomics: How to Fix the Climate and Charge It to OPEC
Steven Stoft with assistance from Dan Kirshner

Copyright © 2008 by Steven Stoft

Diamond Press
4 Roberts Lane
Nantucket, Massachusetts 02554
(508) 825-2195
(510) 644-9410
On the Web: stoft.com

Copy editor: Kathleen Christensen

ISBN 13: 978-0-9818775-0-1
ISBN 10:   0-9818775-0-8

9 8 7 6 5 4 3 2

Printed and bound in the United States of America by Malloy

Contents

Preface

When I started high school, my father gave me a book by a world-class physicist, George Gamow. *Mr. Tompkins in Wonderland* explained quantum mechanics and relativity at a popular level. Travel fast enough for long enough, and you will be only thirty when your twin turns sixty. Some infinities are bigger than other infinities. Contrary to what Euclid said, the three angles of a straight-sided triangle do not always total 180 degrees. The world is full of surprises, and I loved it. I soon discovered that physicists have a tradition of explaining advanced ideas to the public just because they find the concepts fascinating.

Economics, though still a primitive science, contains a few surprising and delightful ideas of its own. Unfortunately, economists seem less interested in explaining their ideas to a broad audience simply because the ideas are fascinating. But I see signs of change, and this book joins what I hope is the start of a flood of popular books about economics.

However, a more practical idea motivated my writing this book. Our nation, and in fact most of the world, is putting in place an enormous and untested set of economic policies and is at risk of a global policy meltdown. Such a failure could waste most of the money we spend and fail as well to achieve its twin goals of climate stability and energy security. With this book, I hope to make such a failure slightly less likely.

My hopes would be higher—but I've been down this path once before. I had the privilege of watching, from up close, the restructuring of the California electricity market—a well-intentioned energy policy with unintended consequences. Later, I acted as the expert witness in the field of economics for the California Public Utilities Commission and Electricity Oversight Board when California sued the Federal Energy Regulatory Commission (FERC) to undo some of the long-term electricity contracts that the state had signed at the height of the 2001 electricity crisis. The state bought $40 billion of electricity, for the distant future, at double the normal cost of power to "protect" Californians from presumed astronomical prices in the future. Three months later, the cost of power was back to normal—but not because of those contracts.

Now, you might think California has little to do with national energy policy, but with prices already double, and a month before the state began to overpay, the Democratic chairman of FERC dragged California and some large power sellers into the Republican White House and instructed California to start buying. The disease of misguided energy policy is national in scope and cuts across party lines.

Watching the development and implementation of new global warming policies feels strangely familiar. People have a lot of enthusiasm and some good ideas, but major programs have already gone far astray. European utilities have made tens of billions of dollars in excess profits from free cap-and-trade permits. The United Nations' Clean Development Mechanism is paying fifty times more than it needs to for emission reductions. In 2006, the United States spent $7 billion on subsidies and higher prices for ethanol with the likely result that worldwide carbon emissions increased.

While it's easy to criticize, my interest lies in fixing energy policy. Technically, that's not so difficult. But it will not happen until more people understand the dangers of trillion-dollar policies based on hunches and appreciate the low cost, power, and simplicity of well-designed policies.

The hardest part of learning new ideas is giving up misconceptions. This is true of physics as well. Even Einstein found the uncertainty of quantum mechanics—the next step after relativity—so disconcerting that he never accepted it. As Mark Twain put it, "It ain't what you don't know that gets you into trouble. It's what you know for sure that just ain't so."

In that sense, economics is tougher than physics, because everyone already knows so much economics "for sure." We all go shopping. So everyone knows for sure they understand prices. It's simple. A higher price makes me poorer and the store richer.

True enough, but what few realize is that prices have a hidden talent for making us rich that surpasses even the best new technology. Adam Smith discovered this in the 1700s and was so impressed that he called it the "invisible hand," which back then meant the hand of God. Strangely, even those who are the most pro-market usually don't believe much in prices. If I explain one economic idea in this book, it's that market prices save you money, and subsidies waste your money. There are exceptions to every rule, but understand why this one is usually true, and I'll make you an honorary economist.

So what does that have to do with energy policy? Price confusion is at the heart of today's energy politics. The policy wonks are saying, "Carbon and oil are priced too low. Fix those prices, and the invisible hand will fix our energy problems." The political interest groups are saying, "Yes. Great idea. Let's raise the price by taxing carbon or selling permits. That will bring in hundreds of billions of dollars for subsidizing our pet projects." A few people are just out to collect the subsidies, but confusion over how the invisible hand works is the biggest part of the problem.

If we learn this lesson, we'll reap an unexpected reward. Since subsidies waste money and prices work on their own, we can have all the money back. That's right. Tax energy and mail all the tax revenues back to consumers on an equal-per-person basis, and the invisible hand still works just as well. Economists have understood this for a hundred years. I know it sounds far-fetched, and it actually is a bit tricky to understand—so I wrote this book.

I won't go into the reasons here, but several other surprising ideas are important for putting our energy policy on track, and with global warming and tightening oil supplies, that's more important than ever. Good intentions do not suffice. An enthusiastic start is no guarantee of future success. Dig deeper, and you find things are not as they seem. That is what this book is about. If we want our energy policy to work better than California's electricity market, we had best pay close attention to the way governments and markets really work.

Acknowledgments

When I decided to write this book, I asked my friend Dan Kirshner, a lifelong environmentalist, to be my coauthor. We met years ago in the University of California at Berkeley graduate economics program, and our paths crossed again in California's electricity market design process. He spent years working for the Environmental Defense Fund, so I knew I could count on him to present an environmental perspective that made sense economically. Although he declined to be coauthor, he did read all the chapters, many of which never made it into the book because he turned out to be a staunch defender of the public's right to clarity. He also suggested innumerable improvements to the content and wrote a couple of the more memorable passages. In short, the book you see before you would not exist had it not been for Dan's continual guidance.

Even Dan's help was not enough, so my friend Joanna, writer extraordinaire, pitched in to straighten out the first few chapters. An author, writing instructor, and onetime journalist, she set me on a new course, leaving both me and my readers deeply in her debt. Hugh Biggar, another friend and an environmental journalist, followed that up with a read that produced a host of helpful clarifications. My copy editor, Kathleen Christensen, has been a delight to work with, going beyond the call of duty by carefully considering my arguments from the viewpoint of the broader audience I hope to reach. Finally, my invaluable proofreader, Ann Marie Damian, made quick work of countless errors.

Any remaining errors are those I snuck in after the others had finished. I own the typesetting program, which proved to be too tempting for someone who can't leave well enough alone.

From the start, François Lévêque pushed my thinking forward with tough questions. I also thank him for publishing chapter excerpts on www.energypolicyblog.com. And thanks too to the rest of the gang—Sandy, Sarah, Doug, and Dave—for their suggestions and helpful news notes, and to Raymond O'Mara, whose sense of design gave the cover of this book some pizzazz.

None of this would have been possible without the encouragement, guidance, and assistance of my wife, Pamela. I would especially like to thank her for putting up with a second book. I also owe a debt of gratitude to my mother, who taught me what good writing is and tried valiantly to pass on some of her skill, and to my father, who taught me most of the science I know.

Disclaimers

Experience indicates that when I write that the market is useful, it will be said that I believe markets solve all problems. So let me make a few things clear.

I do **not** believe

- Markets are more important than government.
- Government is more important than markets.
- Global warming will surely bring disaster if untreated.
- There's a moment to lose on global warming.
- Energy security is secondary to global warming.
- Global warming is secondary to energy security.

As the old saying goes, "predicting is difficult—especially about the future." So I do believe we should hedge our bets immediately.

Acronyms

DOE   The U.S. Department of Energy
CO_2 Carbon dioxide
IPCC   The U.N.'s Intergovernmental Panel on Climate Change
OPEC   The Organization of Petroleum Exporting Countries
GDP   Gross Domestic Product

Numbers

Energy Price Increases Caused by Carbon Pricing

CO_2 $ / ton	Gasoline ¢ / gallon	Oil $ / barrel	Electricity ¢ / kWh
$10	10¢	$5	1.2¢
$30	30¢	$14	3.6¢
$90	90¢	$43	11.0¢

Values rounded for ease of use. Residential electricity costs about 10 cents per kilowatt-hour in 2008.

Throughout the book, carbon prices are given in dollars per ton of CO_2.

Asterisks (*) in the main text indicate material discussed in endnotes.

Part 1

Fossil-Fuel Myths

chapter 1

Once upon a Time

The Stone Age came to an end not for a lack of stones, and the oil age will end, but not for a lack of oil.

—Yamani

ONCE, MANY YEARS AGO in a distant land, Yamani the Enigmatic launched a great experiment. Without warning, he sent out a proclamation to every corner of the earth declaring the need to conserve energy. At first, people conserved little. But gradually, the pace quickened—only to slacken once again.

After six years and only modest progress, Yamani issued a second, stronger proclamation. This time, the world reacted dramatically. For the next six years, while the people of the earth multiplied and grew richer, their use of oil diminished—something never seen before. After twelve years, Yamani and his confederates, duly impressed with the power of their methods and the world's response, withdrew their proclamations.

There matters rested for another eighteen years. Surprisingly, much of the world's reaction continued, and by the end of the thirty-year experiment, the world had saved, by a most conservative estimate, eight times as much oil as it now uses in a year.*

The story is true. Yamani has retired, but his confederates have begun a second and more sophisticated experiment. Fortunately, the lessons of that first experiment, if properly applied, provide a path to escape the enormous costs that now await us if we fail to choose a secure and sustainable energy future.

Sheikh Ahmed Zaki Yamani, famous for his enigmatic sayings, was Saudi Arabia's oil minister when OPEC, the Organization of Petroleum Exporting Countries, conducted its "great energy experiment."[1] The first "proclamation" led to the October 1973 oil shock, which tripled the price of oil. The second "proclamation" led to the 1979 oil shock, which doubled the price again.

While the worldwide response was enormous, the U.S. response was even more dramatic. U.S. addiction to oil decreased over a thirteen-year period, as did the country's carbon dioxide (CO_2) emissions. The United States conserved not just oil, but all kinds of energy. In the thirty years from 1973 through 2003, the United States saved energy equivalent to twenty years of U.S. oil consumption at the rate we now consume it.*

Carbonomics, the economics of fossil fuels, not only explains that astounding success, but also teaches us how to repeat it—but this time without paying OPEC another trillion dollars in tribute.

Yamani's experiment did more to reduce CO_2 emissions than the Kyoto Protocol has; there is simply no comparison. The experiment taught the world how to gain independence by saving energy, how to stabilize the climate by saving carbon, and how to increase security by reducing the world price of oil. By 1986, these lessons were fairly well understood, but OPEC had been crippled, and climate change was not yet a concern, so there was little motivation to act on the new understanding. As a result, nothing was done, and now the lessons are forgotten.

Climate Stability and Energy Security

The key to an effective energy policy is to understand that climate stability and energy security are twin challenges—though not identical. Both are global issues, and both suffer from the problem of free riders, which I describe later in this chapter. Unfortunately, those interested in one challenge often show little interest in—and sometimes antagonism toward—the other. I believe the two challenges—climate stability and energy security—are not only compatible, but that solving either requires solving both.

Twin Global Challenges. It's clear that global warming requires a global solution, but Yamani's experiment taught us that energy security also requires a global solution. In 1974, the United States recognized the need for a global response to OPEC, and Secretary of State Henry Kissinger organized what the *New York Times* called "a counter-cartel of the major oil-consuming countries." That organization, the International Energy Agency (IEA), still exists; twenty-seven countries including the United States, Japan, and most of Europe are members. But it has forgotten its purpose.

---

1. You can meet Yamani at www.azylawfirm.com/founder.asp.

Later, in 1979, after OPEC doubled the oil prices that it had already tripled, the seven industrialized nations held a "world economic summit." They issued a communiqué, which the *New York Times* again said "amounts to a consumers' cartel." This effort also failed; nevertheless, the global response to high oil prices eventually did crush OPEC—but not permanently.

Now, the lessons that Yamani's experiment taught have been forgotten, and people think the United States can achieve energy security on its own. But even if Americans cut oil imports to zero—say, by driving hybrid cars that burn ethanol—we would not achieve independence. The world oil market would still control the price of corn ethanol at American gas pumps, just as it does now. I explain this in Part 2, along with other lessons, including how to crush OPEC again.

So energy security is a global challenge just like climate stability. OPEC's market share has grown again, and OPEC is short on production capacity, as it was before 1973. China and India are rapidly expanding their demand for oil. Greenhouse gas emissions are increasing faster than ever, and China has passed the United States to become the largest emitter of CO_2. No one country, not even the United States, can meet either challenge on its own.

The Problem of Free Riders. By curbing our use of oil, we can force down its price on the world market. While this is worth doing, the job is tough if we go it alone. Any price decrease we cause benefits all consumers worldwide, even if they do nothing to help out. Economists call those who benefit without helping out "free riders." These free riders take advantage of the lower price to use more oil, counteracting our efforts.

Climate change presents a parallel problem. No country, acting alone, can do much to stop climate change. Any country that tries will find that most of the benefit accrues to other countries. So the more we do to reduce global warming on our own, the less others will worry about global warming, and the less they are likely to help.

Solving the problem of free riders requires an international approach, such as the Kyoto Protocol. But energy security also requires a global approach—a point that Kissinger's team recognized in 1974, but which is now forgotten. Fortunately, because the challenges are twins, the same international organization can address both. But we need a better design than the Kyoto Protocol or the IEA offers. Part 4 provides a blueprint of that better design along with the rationale for unifying these two problems and their solution.

Conflicting versus Joint Solutions. Some proposed solutions that help with one challenge conflict with the other. Joint solutions, however, help us meet both challenges. One proposal for increasing energy independence conflicts most intensely with solving the problem of climate change: turning coal into gasoline. Unfortunately, this proposal is a favorite of Big Oil and Big Coal.

Coal companies like the idea of making gasoline from coal for obvious reasons—it takes a lot of coal. But oil companies are just as enthusiastic because they would build and operate the new coal refineries. The problem is, these refineries use far more fossil energy than oil refineries do, which is terrible from a global warming perspective.

Fortunately, conservation—the main activity that crushed OPEC in the early 1980s—is an ideal solution, though not the only solution, for both challenges. In fact, conservation is better for energy security than producing gasoline from coal. Of course, the oil companies hate conservation, which is shorthand for not using their product. Gasoline made from coal keeps us addicted and keeps us paying prices set by the world oil market. Conservation helps us break the habit.

Cooperation

Breaking the world's oil and coal habits is no easy task, and those who think it can be done by either resolute proclamation or a change of consciousness will once again be disillusioned. Only a program with the broadest support and based on self-interest can succeed. This explains why joint solutions are crucially important. Only joint solutions can provide the basis for broad-based national and international cooperation.

National Cooperation. The chance of achieving a sound energy policy is now better than ever, because we have a double motivation. OPEC is again breathing down our necks, and climate change has become the number-one national concern on the environmental front. But Americans divide into two camps over which challenge deserves priority. One camp focuses on energy security and the other on climate stability. If one camp adopts a policy that conflicts with the goal of the other camp, the double motivation is lost; in fact, the two camps could cancel each other out.

On the other hand, adopting a cooperative strategy could produce a complementary alliance between the two groups. The environmental camp can provide the staying power and the link to popular international concern about energy issues. The energy security camp can provide the motivation that comes from the short-term tangible gain that is possible in the oil market. It took only six years to bring about a huge reduction in world oil prices after OPEC doubled oil prices in 1979 and 1980. It will take much longer to have an impact on climate change.

International Cooperation. China and the United States together emit half of all greenhouse gases, yet neither has made a commitment to take specific action. If these two countries fail to cooperate, the world has no real hope of success against global warming. And nothing substantial will be done about OPEC's increasing market power and the tightening oil market.

Although both countries claim to be concerned about global warming, both are also afraid of reducing economic growth. As things now stand, neither is likely to make or keep a strong commitment.

One thing, however, could motivate China and the United States to come together. Both are addicted to oil, and their addiction is growing. China is predicted to increase its oil imports from 20 percent of the country's oil use now to about 80 percent in 2030. China is already building plants to refine coal into gasoline. Any reduction in the world price of oil would provide a huge economic benefit to both countries. Surprisingly, only one thing is likely to lower global oil prices—an effective international climate agreement.

An international climate agreement is also, like it or not, an oil consumers' cartel. A consumers' cartel is simply an international agreement to use less oil, and any effective climate agreement will make sure we do just that. Instead of hiding this fact to avoid upsetting OPEC, we should advertise it to enhance the appeal of an international agreement.

That a climate agreement is automatically an oil consumers' cartel may come as a surprise, but it shouldn't. Among economists it's an open secret. In fact, in 1998, when the U.S. Department of Energy (DOE) analyzed possible U.S. compliance with the Kyoto Protocol, it found that even such a weak agreement would have served as an oil consumers' cartel—though it did not use the word *cartel*. The DOE found that the Kyoto Protocol would have lowered the world price of oil by 16 percent had the United States fully complied. With oil at $100 a barrel, that would have saved the United States $70 billion a year on imported oil. American consumers—who must pay domestic oil companies as well as OPEC—would have saved over $100 billion a year.

Unfortunately, the Kyoto Protocol is fatally flawed. It does not require developing countries to make any firm commitments to reducing emissions. This is one reason the U.S. Senate voted against such a treaty 95-to-0. Our problem with the Kyoto oil consumers' cartel—if I may call it that—is much the same problem that Yamani had with the OPEC cartel. Smaller OPEC producers went for a free ride at Saudi Arabia's expense. They did not restrain their production, leaving that job to Yamani.

Developing countries take a free ride on the Kyoto Protocol by not restraining their consumption. This damages both climate stability and energy security.

Although our organizational problems are similar to Yamani's, a consumers' cartel has two organizational advantages over OPEC. First, the consumers' cartel can piggyback on the goodwill and momentum of international climate initiatives. Second, according to experts in the field, a climate agreement can use international trade law as an enforcement mechanism.

The oil price benefits of an international consumers' cartel do not detract from its climate stability benefits. The two are entirely complementary. In fact, to garner support, the proponents of any climate agreement need to take advantage of people's short-term self-interest, playing up the five years it takes to reduce oil prices, as opposed to the fifty or so years it could take to solve the problem of global warming.

Part 4 of this book discusses how to put together a durable international organization that challenges OPEC and stabilizes the climate. The first step is to replace the emissions-cap policy that has stymied the Kyoto Protocol. The second step is to use China's interest and the U.S. interest in lower oil prices to lever these two into an international agreement with binding commitments. The third step is to curb the problem of free rides with an enforcement mechanism better than anything Yamani ever dreamed of.

None of these ideas are new. For example, the move away from international carbon caps has support from a wide range of experts, from George W. Bush's chief economist N. Gregory Mankiw, to liberal economist Joseph E. Stiglitz. But the ideas are important because the people currently debating national energy policy are ignoring these important international considerations and may well end up obstructing rather than advancing international cooperation.

A Fossil Philosophy

So far, I've mentioned the twin challenges, joint solutions, learning from OPEC, and free riders. Another theme of this book is prices and markets. Most people consider pricing to be weak medicine compared with government mandates such as a strict cap on carbon emissions. But markets—driven by prices, not mandates—have built the modern world, with its engines that consume 40,000 gallons of oil per second (this is not a typo). If prices are strong enough to drive the world's economies, they are strong enough to meet our present challenges.

Another theme of this book is conservation, which many also consider weak medicine. Conservation, however, moved quickly and vigorously against OPEC. In fact, it moved ten times more forcefully than all the increases in energy supply—including non-OPEC oil supplies, nuclear energy, and synfuels.

Just a few ideas underlie all of the themes of this book. These ideas make up a sort of fossil philosophy. As with all philosophies, we cannot follow this one to the letter. But it does provide guidance in many situations. The underlying ideas are these:

- ► Treat the problem, not the symptom.
- ► Support cooperation.
- ► Focus on real benefits, not imaginary disasters.

These are the simple tenets that guide the energy policies of this book. But simple as they are, they are often forgotten.

Treating the problem instead of the symptom is the most important. We rely too much on coal and oil and not enough on wind and conservation. Those are the symptoms. But why do we do that? What is the underlying problem?

The price of oil does not include the military cost of protecting oil supplies or the cost of oil's effect on the climate. So the price of oil has long been too low. That is the root problem. Not having enough wind turbines is only one of a million other symptoms, large and small. Using the government to try and fix a million symptoms is, according to the first principle of fossil philosophy, a bad idea.

Of course, the first principle wouldn't be worth much if a million underlying problems led to the million symptoms. But, in fact, only four major problems account for almost all the symptoms. Called market failures, the four underlying problems are these:

- ► The low price of carbon (fossil fuel).
- ► OPEC's market power.
- ► The nearsightedness of consumers.
- ► Insufficient reward for advanced research.

Not that it will be easy, but fixing these four problems is all we need to do to meet the twin energy challenges.

Pricing Carbon. We can raise the price of carbon with either a cap-and-trade policy, a tax on carbon, or an untax on carbon. A central purpose of this book is to explain the old and venerable concept of an untax. The term is mine, and I hesitate to introduce it. But the economic description—"a Pigovian tax with a full, equal-per-person refund"—seems a bit awkward. In any case, the untax is a combination of a carbon tax and a per-person refund that the government mails out, say, once a year. An example is Alaska's Permanent Fund, which issues annual refunds of revenues from the Trans Alaska oil pipeline to Alaska's residents.

While refunding a tax may seem circular, the untax provides more bang for our bucks than any other energy policy. I explain this economic mystery in Part 3, but for now I will simply note that, in July 2008, Al Gore called an almost identical proposal "the single most important policy change we can make." But this is no liberal nostrum. Former Bush economist Mankiw supports a proposal identical to Gore's, and the right-wing American Enterprise Institute is on board. James E. Hansen, the most outspoken climate scientist, also proposes an untax by a different name.

A Consumers' Cartel. The solution to the second problem—OPEC's market power—is, as I've already mentioned, an international consumers' cartel. Although, in 2007 and 2008, Saudi Arabia was withholding about 20 percent of its oil production capacity and has underinvested in new capacity for twenty years, OPEC may not be the main supply problem. The main problem might be natural limits—that is, we might be near the peak of cheap conventional oil production. Fortunately, a consumers' cartel works even better against a natural shortage than against an antagonistic producers' cartel.

A Race to Fuel Economy. When making purchases that can save energy over many years—for example, the purchase of a house or car—consumers tend to be systematically nearsighted. That is, they undervalue future energy savings. So consumers don't push automakers as much as they could to improve fuel efficiency. We can address this failure of the energy market by engaging car companies in a race to produce fuel-efficient cars. This eliminates the need for government standards and produces a more powerful incentive to improve.

An Energy Moonshot. Lately, people have been talking about the possibility of an energy moonshot—a major effort something like Project Apollo, which put a man on the moon. This could correct the fourth market failure, a shortage of funding for advanced research. However, we need to be careful. This market failure justifies government funding of basic research but not vast subsidies for existing technologies. Clean coal technology is an excellent example of an energy moonshot the government should fund.

⌒

Most of Part 1 concerns energy myths. In Chapters 2 through 4, I demonstrate the importance of rejecting imaginary or speculative disasters. To balance things out, I debunk the myth of energy miracles in Chapter 5, while in Chapter 6 I question the most pessimistic view of policy. And for those anxious for answers, the last chapter of Part 1 summarizes the national policies that I propose in more detail in Part 3.

However, before I go into detail about my proposals, I lay a foundation in Part 2 for understanding both national and international policies. In Parts 3 and 4, I focus on solutions to the four basic failures of the energy market. Parts 3 and 4 also focus on cooperation, at both the national and international levels—even down to the level of car companies.

I have designed this book to help readers who wish to skip ahead. But for a solid understanding of why the policies I propose are necessary and cost-effective, I suggest you first clear your mind of the myths about fossil fuel right here in Part 1, then read about the realities of energy markets in Part 2.

chapter 2

Wreck the Economy?

The Kyoto treaty would have wrecked our economy, if I can be blunt.

—President George W. Bush, 2005

IF I MAY BE BLUNT MYSELF, of all the fears concerning climate change and addiction to oil, the fear of wrecking our economy is most paralyzing but least substantial. Even if the costs were greater than they actually are, for America to turn away in fear from the challenges of climate and addiction would dishonor our heritage and lay our own responsibilities at the feet of future generations.

The irony of America's recent energy policy is that, by taking little responsibility for our energy use, we have once again handed the power of the oil market to the Organization of Petroleum Exporting Countries (OPEC). The connection is straightforward. The Kyoto Protocol calls on nations to reduce their use of fossil fuel, mainly coal and oil. Reducing the use of oil makes oil less scarce and reduces its price. In fact, as I mention in the previous chapter, a reduction in the world's use of oil was what crushed OPEC's market power for eighteen years.

Our choice is not between a wrecked economy and economic growth. It is between controlling our own energy policy and letting OPEC's high prices force upon us an energy policy of its own design. Theirs is a poor policy indeed, as OPEC profits from our addiction and dislikes policies that stop global warming. But its policy is forcing us to conserve oil. By 2007, our rising oil use leveled off, and in the first half of 2008 U.S. oil use was down over 2 percent from a year earlier and oil imports were down 2.5 percent. Compare this with an annual growth rate in oil use of 1.5 percent in the decade before 2005. President George W. Bush claims credit for reducing energy intensity—energy use compared with gross domestic product (GDP). But the reality is that OPEC's high prices are making us conserve—just as they did in the 1980s—while the economy continues to grow. While conservation is a benefit, when administered by OPEC, it comes at far too high a price.

Instead of idly waiting to see what OPEC had in store for us, we could have chosen our own destiny. Our own market-based policies could have guided the use of better technology to reduce our dependence on coal and oil. According to the Department of Energy (DOE), this would have reduced the world price of oil—just as it did in the 1980s. The DOE discovered this in 1998 when Congress asked it how signing on to the Kyoto treaty would affect our economy. The DOE also discovered that implementing the Kyoto Protocol, flawed as it was, would not wreck our economy.

It is too late to avoid paying the present round of tribute to those powers both foreign and domestic that control the world's oil. But we can, in a few years, regain control of our energy destiny by heeding the advice of a president who presided over some of the most perilous times in U.S. history. Even before confronting the perils of World War II, Franklin D. Roosevelt faced the dangers of the Great Depression. He did not flinch, saying, "Only a foolish optimist can deny the dark realities of the moment." But he also warned of the greater danger of being ruled—and paralyzed—by fear, famously declaring "We have nothing to fear but fear itself."

Just as it was seventy-odd years ago, fear itself is again our greatest enemy. That's why I begin this book by dispensing with the exaggerated predictions of economic ruin, catastrophic shortages, and unstoppable climate change. And although the book is motivated by the real dangers of global warming and the dependence on foreign oil, I do not dwell on these. Instead, I present a plan to improve our chances against both threats, without wasting money and at a surprisingly low cost. Although no panacea exists, what we need as a nation is courage, cool heads, and a clever, low-risk plan of action.

Overcoming Fear

Only after we lay to rest the fear of taking action will it make sense to plan a more secure and environmentally sound energy future. But after so much misleading rhetoric, a simple claim that the U.S. economy is strong will not suffice. The belief in economic damage is so ingrained that it afflicts even some of those most willing to take action.

Undoing those misconceptions requires looking at energy policy from all angles—from the expert, rather than the political, perspective; from the perspective of economic growth; from the perspective of physical possibility; and, finally, from the present perspective of inaction.

To begin, consider what the government found out when it studied the cost of complying with the Kyoto Protocol. In 1998, Congress asked the DOE to examine this cost. Congress required the DOE to assume that we would begin complying as late as possible and then comply suddenly. Congress also prohibited analysis of fuel-economy or energy-efficiency standards. It allowed the DOE to model only a carbon tax.

In spite of those cost-increasing assumptions, the DOE found no reduction in long-term economic growth. It found that the shock of sudden compliance would cause a temporary slowing of growth. But the report predicted that, by 2020, our gross domestic product (GDP) would be less than 1 percent behind the no-Kyoto scenario.

But what about more-recent proposals that seek to accomplish even more than the Kyoto Protocol does? For over twenty years, economists have been estimating the costs of energy policies. Researchers have performed dozens of such studies and have generally found costs in the range of 1 to 3 percent of GDP for strong policies. I will use a cost of 2 percent as a benchmark, though most proposals predict that costs will increase slowly, not reaching 2 percent for decades. I will return to the question of why the cost is so low after I dispense with a more urgent question.

Could a 2 Percent Cost Stop Economic Growth?

Confusingly, politicians and pundits always seem to tie energy program costs to reduced economic growth. This happens so consistently that when I first checked on costs, I was afraid that an effective policy would reduce the economy's growth rate by 2 percent—from a normal 3 percent per year to 1 percent per year. That would indeed wreck the economy.

When President Bush announced his Global Climate Change Initiative on Valentine's Day 2002, he said: "Our nation must have economic growth—growth to create opportunity; growth to create a higher quality of life for our citizens. Growth is also what pays for investments in clean technologies, increased conservation, and energy efficiency."

It sounds as if growth itself is in question. Perhaps if Bush had picked the wrong climate-change initiative, the United States would have stoppped growing. This didn't sound right to me. But if it were true and the country grew even 1 percent slower for 100 years, the economy would make almost two-thirds less progress. Such a dire outcome worried me, even though the no-growth rhetoric appeared to be based on pop economic theory or on a misunderstanding of real economics. The administration cited no studies or papers to support its dire predictions.

For help with this question, I turned to the work of Dale W. Jorgenson. Jorgenson has a chair at Harvard, has been president of the American Economic Association, and has won many honors in economics. Perhaps more important, he is the man who wrote the book, figuratively and literally, in this area of economics. So I bought Jorgenson's *Growth: Energy, the Environment, and Economic Growth*, volume 2.

The first study in the book analyzes the OPEC crisis of 1973 to 1986, the original great energy policy "experiment." Of all the studies estimating the costs of an economy-wide policy, this one appears to be the most reliable, because it examines a policy experiment—OPEC's—that was actually carried out. Most studies examine proposed future policies. The strength of the OPEC policy provided Jorgenson with an ideal data set for his analysis.

Two of his most interesting scenarios he calls OIL72 and OIL81. The first represents what would have happened if OPEC had never raised the price of oil higher than $12.50 per barrel (in 2007 dollars), the price in 1972. The OIL81 scenario represents what would have happened if the oil price had stayed at its 1981 value of about $90 per barrel. In the first scenario, the country would have been a bit richer, and in the second scenario a bit poorer. The difference is equivalent to a policy that raises the oil price from $12.50 to $90 and keeps it there permanently. Jorgenson found that such a policy would have reduced GDP by 2.5 percent.

That's 2.5 percent total in the long run—not 2.5 percent per year!

Jorgenson's analysis shows that ten, twenty, or a hundred years after oil reached $90 per barrel, the United States would be 2.5 percent poorer than if oil had stayed at $12.50 per barrel. This tells us that growth has not slowed down permanently. After a one-time reduction in GDP, full-speed growth would resume. If growth had slowed permanently, GDP would have fallen further and further behind each year

Although this is probably the most convincing analysis, because it is based on a wealth of real-world data and examines a harsh policy, Jorgenson's analysis is completely in line with every analysis of long-term economic growth that I have examined. An energy policy that makes a large, fixed, and permanent increase in the cost of fossil energy causes a small initial reduction in growth, but then growth resumes at full speed forever after.

This does not surprise economists. Technological progress is the main determinant of long-term growth, and energy policy does not slow technological progress. In 1997, over 2,600 economists—including nine recipients of the Nobel Memorial Prize in Economic Sciences—signed the Economists' Statement on Climate Change, which concludes:

> For the United States in particular, sound economic analysis shows
> that there are policy options that would slow climate change *without
> harming American living standards*, and these measures may in fact
> improve U.S. productivity in the longer run [emphasis added].*

In other words, economists do not believe the wreck-the-economy myth. They
believe that many potential policies could reduce greenhouse gas emissions and
not harm—let alone wreck—the American standard of living. In fact, economists
believe those policies might actually improve productivity.

Looking at the historical performance of the U.S. economy tends to confirm this finding. In 1982, the economy slumped, but in the next three years
it grew 4.5 percent, 7.2 percent, and 4.1 percent—quite a record, considering
average growth is only about 3 percent annually. And all the while, OPEC was
imposing its superaggressive climate policy—to put it charitably.

So that answers this section's question. A policy that costs 2 percent
of GDP does not wreck economic growth. Imposing a 2 percent cost on the
economy slows its growth only until the GDP has fallen 2 percent behind. After
that, growth resumes at its full normal rate. Think of it like this: If I have to give
up my two SUVs for hybrids, I might be 1 percent poorer now, and I would
still be 1 percent poorer in ten years. But I won't be 10 percent poorer after ten
years. Once I make the switch, my income resumes its normal growth.

Is 2 Percent a Large Sacrifice?

President Richard Nixon announced Project Independence just three weeks after
the start of the oil embargo in 1973, when Arab nations stopped shipping oil to
countries that supported Israel in the Yom Kippur War. "We must ask everyone
to lower the thermostat in your home by at least six degrees," said Nixon, "so
that we can achieve a national daytime average of 68 degrees." President Jimmy
Carter endorsed the same temperature and suggested wearing a sweater.

But over the past thirty years, the talk of sacrifice has shifted dramatically.
Even among environmentalists, only a few emphasize sacrifice, and most don't
think much sacrifice is necessary. New Mexico governor Bill Richardson, in
an interview posted on the online environmental publication Grist, expresses
the current view most clearly:

> I believe it's going to take … sacrifice for the common good.
> … What I'm asking for is not sacrifice, like Americans wearing
> sweaters and turning the heat down. What I'm asking for is being
> more energy-efficient with appliances, with vehicles, with mass
> transit. Maybe, instead of driving to work, once a month go mass
> transit.

Richardson is not wrong, but he's missing a crucial part of the picture. Usually, sacrifice means getting by with less. A strong energy policy does not require that. It costs us something, but even with that "sacrifice" we will get by with more, not less. But we won't have quite as much more as we could have had.

Here's an example of "sacrifice" with growth. The Apollo program sent a man to the moon but made us poorer than we would have been—that is, we paid extra taxes to cover the program's cost. But it didn't hurt our economic growth rate. The United States grew richer at the same time as Apollo's costs were increasing. The costs increased more slowly than the economy grew, so the "sacrifice" for Apollo didn't actually make the country poorer. On the day we landed a man on the moon, the country was richer than on the day President John F. Kennedy announced that goal—just not quite as much richer as it might have been.

Perhaps it's worth restating the obvious at this point. The purpose of an Apollo program or an energy program is to buy a moon landing, a better climate, or increased security. That's why there is a cost. If the policy is wise, the benefit will outweigh the cost. The gain will be worth more than the sacrifice. In any case, the cost does not slow economic growth; it just takes a bite out of our income.

In April 2007, researchers at the Massachusetts Institute of Technology (MIT) looked forty years into the future at the impacts of seven cap-and-trade bills before Congress. Each would place a decreasing cap on greenhouse gas emissions. Figure 1 shows the increase in consumption per person (not per family) from 2010 until 2050, under the strictest scenario modeled by the MIT group.

Consumption of goods and services more than doubles, from $31,900 per person in 2010 to $74,500 per person in 2050. But with a strict greenhouse gas policy, consumption is 2.4 percent less in 2050 than without the policy. The "sacrifice" means getting 128 percent richer instead of 133 percent richer.

The "sacrifice" is relatively small in the first few years under the strict policy. After ten years, consumption is only half a percent lower than it would have been. The policy requires deeper cuts in CO_2 over time—about 50 percent after fifteen years, relative to a case in which no policy is in place, and about 75 percent after forty years.

The economy fall further behind over time not because economic growth is damaged, but because the policy becomes stricter. If energy problems abate and the policy does not require further lowering of the cap, the rate of economic growth is unaffected. A policy with an unchanging cap has no impact on growth.

Figure 1. Effect on Personal Consumption of a Strong Cap-and-Trade Policy

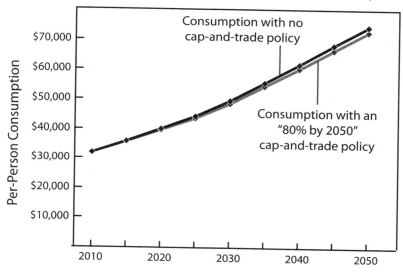

As the cap of a cap-and-trade program tightens, it takes an increasing bite out of income. This graph assumes a program that targets an 80 percent reduction in greenhouse gas emissions by 2050 and is similar to programs proposed by Congress. After forty years, average per-person (not per-family) consumption would have reached only $72,700 instead of $74,500. The graph is based on data published in April 2007 by a team of researchers at MIT.*

Another way to think of the "sacrifice" required is as a delayed increase in income. Under the strict policy that the MIT team studied, the country must wait until 2051 to achieve the income it could have attained in 2050.

How Can It Be So Cheap?

You may now be wondering if the economists who come up with these numbers are in touch with reality. How could it be so inexpensive to cut back on fossil fuel, the very lifeblood of a modern economy? Why are we so addicted if it's so cheap to switch?

The basic answer is this: The United States is rich, and fossil fuel is not as costly as you might think. In fact, it has been too cheap to pass up. Much of the cost of electricity and gasoline is not the cost of fossil fuel, but of wires, generators, and refineries.

The DOE's 1998 model predicted that the largest carbon savings would come from replacing coal-fired generators with natural-gas-fired generators. Coal is higher in carbon per unit of energy produced than other fossil fuels and produces 35 percent of U.S. CO_2 emissions. Natural gas is the cleanest fossil fuel and generates electricity more efficiently. So using gas instead of coal would reduce U.S. CO_2 emissions by about 20 percent, a good start. How much would that cost us?

Coal is cheap. All the coal we use costs only 0.2 percent of GDP. That's two one-thousandths of domestic production. However, coal plants are more expensive to build than gas-fired plants and are less efficient. So although gas cost several times more than coal per unit of energy, electricity produced with gas is not that much more expensive. Switching from coal- to gas-fired power plants would increase electricity costs only about 2 cents per kilowatt hour. (The retail price is about 10 cents per kilowatt hour.) This would cost about $40 billion dollars a year or 0.3 percent of GDP. If enough other fixes could be found that were equally cheap, fossil CO_2 emissions could be eliminated completely for a cost of 1.5 percent of GDP.

Wind power is a little more expensive than electricity from natural gas, but it has the potential to eliminate 100 percent of CO_2 emissions. So it's almost as cheap a way to reduce emissions as switching to natural gas. A third option is nuclear power. It costs about the same as wind power and also eliminates CO_2 emissions. As an aside, building power plants of any kind emits some CO_2, but the amount is very small compared with the amount emitted by producing power with coal.

What about oil? When oil costs over $100 a barrel, I cannot escape a startling conclusion. OPEC and the world oil market have already forced an oil conservation policy on us, in the form of high oil prices. This "policy" is as costly as the oil component of the strictest actual climate-change policy. We do not need to spend more than we are already spending. Instead, we need to take those revenues back from OPEC and Exxon and use them to implement a real policy that is just as effective as OPEC's unofficial one.

High oil prices have, for three years running, stopped the growth in oil use, and even initiated a decline. As long as oil costs over $100 a barrel, we can reduce emissions as much as we need to at no extra cost for a couple decades.

Even Cheaper?

The cost of alternative energy is easier to pin down than the cost of conservation, so I use alternative energy as a reliable way to show that the costs of reducing greenhouse gas emissions could be low. But when OPEC raises prices, the world

No Guarantee

Economic estimates of low cost are not a guarantee. You can buy a cheap used car, but buy a lemon and repairs can triple the cost you expected.

Most economic estimates assume, as the DOE did, the use of an energy policy similar to the one I recommend. But adopt a huge ethanol program or mandate an end to fossil-fired electricity in ten years, and all bets are off.

Non-market-based programs could easily cost ten times more than expected or not work at all.

responds mostly by conserving, because more cheap conservation is available than cheap alternative fuel.

A 2007 report from McKinsey and Company, the world's leading management-consulting firm, examined dozens of approaches to abating greenhouse gases, including conservation measures, forestation, and alternative fuels. The company found that the world can accomplish a large fraction of the required emission reduction at a cost savings (a negative cost) of half a percent of world GDP. For example, better insulation can save more by reducing oil and gas costs than it costs to insulate. To be cautious, the authors of the report count the negative cost as a cost of zero, then double their total estimated cost. The report concludes that an aggressive policy could cost 1.4 percent of world GDP.[1]

Taking Charge of Oil Policy

How did OPEC regain its power? Before the 1973 oil embargo, the United States spent under 2 percent of its GDP on oil. Then, for a few years, it spent 5 to 6 percent. In 1979, the cost spiked to 9.9 percent, and the world began to take oil prices seriously. By the end of 1985, worldwide conservation had crushed OPEC, and for eighteen years—until 2004—the United States again spent, on average, under 2 percent of GDP on oil.

During the eighteen-year grace period, and especially in 1986, people had two points of view. Some said to keep the price high so we would keep conserving and keep OPEC at bay. Others said they liked the low prices. "Liking low prices" won out.

Keeping prices low had the predicted effect. Conservation partly petered out, and the much-smaller increase in oil supply petered out completely. Meanwhile, OPEC wisely stopped the growth of their production capacity and waited for world oil use to grow. It has grown, and prices went back up. With oil at $100 a barrel and with GDP at the 2008 level, the United States spends 5.5 percent of GDP on oil, up from 1.7 percent in 2002. OPEC's recent "energy policy" is a lot like a Kyoto policy focused on oil, but with a startling difference.

> **The DOE's Conclusion: Kyoto Would Cut the Price of Oil**
>
> In its 1998 report on the effects of the Kyoto Protocol, here's what the DOE predicted: "Because of lower petroleum demand in the United States and in other developed countries that are committed to reducing emissions under the Kyoto Protocol, world oil prices are lower by between 4 and 16 percent in 2010, relative to the reference case price of $20.77 per barrel."
>
> The 16 percent value is based on full compliance, and the 4 to 16 percent range in oil price reduction indicates that U.S. compliance would have the dominant effect on world oil prices under the Kyoto Protocol. (Lack of U.S. compliance nearly eliminates the oil price reduction.)

---

1. This is their cost estimate for a policy that would "cap the long-term concentration of greenhouse gases in the atmosphere at 450 parts per million (ppm)." We are now just over 380 ppm.

In 1998, the DOE concluded that the United States, to comply with the Kyoto Protocol, would need to push the price of gasoline up to $2.31 per gallon (in 2007 dollars). Similarly, the MIT researchers found that a price of $101 per barrel of oil was sufficient up through 2030. In other words, in mid-2008, oil and gas cost more than enough, and much more than was expected from compliance with the Kyoto Protocol.

But that's not the difference I'm talking about. To see the real difference, follow the money. The DOE assumed that the government would refund revenues from the tax on oil "to consumers through a personal income tax lump sum rebate." In other words, all the higher gas costs of a Kyoto policy could have been returned to you and me in the form of annual checks from the government. (I will explain how this works in Chapter 7.) That's the way Alaska returns revenues from its oil pipeline to its citizens. Needless to say, when OPEC and Exxon raise the price of gasoline, they forget to put the check in the mail. That's the enormous difference between implementing our own policy to comply with Kyoto and letting OPEC impose a policy on us.

There is no doubt that paying OPEC is worse than paying ourselves, but with a Kyoto-style policy, wouldn't we have had to pay both at once? The answer is no, for two reasons. First, gasoline prices need to be only so high to encourage conservation—say, $3.50 per gallon. To the extent OPEC raises the price, we don't need to. Second, if we raise the price of oil before OPEC does, that curbs oil use and makes it harder for OPEC to raise its price.

Had we implemented a Kyoto policy in 1998, we would have preempted OPEC by six years. The DOE estimated that a Kyoto policy could have cut OPEC's prices by 16 percent. However, the policy the DOE examined focused on coal and included no fuel-economy measures. With a policy focused more strongly on oil, we could have reduced OPEC's price even more. Also, the DOE report did not anticipate an oil market as tight as it is now. When the market is tight, an oil conservation policy has more impact on price.

The DOE is not alone in predicting that climate and energy independence policies will reduce OPEC's price. For example, the MIT climate-policy model predicts a 47 percent reduction in the world oil price by 2050, and others have made similar predictions. The idea that reducing demand reduces price dates back to Adam Smith. That's just how markets work—even when a cartel controls part of the market.

An Oil Policy That Works

As I explain in more detail in Chapter 7, a good oil policy includes an untax on oil and a fuel-economy incentive for carmakers. Untax is the term I use for the DOE study's method of refunding all revenues. It's not a tax, because the government keeps none of the revenue. The point to understand here is that the government refunds all revenues on a per-person basis, the way Alaska handles its Permanent Fund.

The untax keeps encouraging us to use less oil, even if OPEC lowers its price. Here's an example: Suppose the price of oil is $100 per barrel when we implement an untax. The starting untax rate is zero, because the oil price is already high enough to encourage consumers to save oil. If the price goes down to $80, the untax goes up to $20 a barrel. For consumers, it's the equivalent of having the price of oil stay at $100. They keep conserving and buying alternative fuel. But consumers still benefit from OPEC's price reduction. The government refunds all the money collected—$20 per barrel on 20 million barrels per day—by sending checks out in June on an equal-per-person basis, just like Alaska.

Keeping the domestic price of oil effectively at, say, $100 per barrel while pocketing the difference between that price and the actual world oil price holds down demand, which holds down the world price of oil. When the price of oil is $100 a barrel or more, we're already paying the most expensive part of a climate policy. As the world oil price comes back down, and we pocket the difference between that price and $100, climate policy will only get cheaper.

In every decade since 1920, U.S. income has increased faster than energy use has. Adjusting for inflation, we now have three times as many dollars to spend per unit of energy that we consume. With a sound energy policy that ends up costing 2 percent of GDP in 2050, the fraction of our income that we spend on energy will continue to decline.

~

The Kyoto Protocol puts no restriction on how countries curb their emissions of greenhouse gases. So when President Bush claimed the treaty would wreck the economy, he was claiming that any serious climate proposal would wreck the economy. In 1998, however, the DOE found that by 2020, more than a decade after its start, the protocol would have reduced GDP by less than 1 percent—not from its 1998 level, but from its predicted level in 2020.

Experts generally estimate that strong greenhouse gas programs cost roughly 2 percent of GDP. Those who claim such programs will wreck the economy generally speak of a reduction in economic growth. And, indeed, a 2 percent reduction in the rate of growth would be devastating.

But climate programs of constant strength cause no reduction in economic growth. They only cause a continuing cost. The difference is enormous. A 2 percent reduction in growth would cut our income in half after forty years, while a 2 percent continuing cost only cuts it by 2 percent.

Every economic analysis of climate proposals points to a continuing cost, not a reduction in the rate of economic growth. This means we might have to wait until 2051 to be as rich as we would otherwise be in 2050—which will be more than twice as rich as we are now. This is the case even with climate programs stronger than the Kyoto Protocol. So do not fear. Market economies are strong and not so easily wrecked.

chapter 3

Peak Oil or Liquid Coal?

Civilization as we know it will come to an end sometime in this century, when the fuel runs out.

—David Goodstein, Professor of Physics, Caltech

PEAK-OIL THEORY COMBINES serious geology with pop economics to "envision a dying civilization, the landscape littered with rusting hulks of useless SUVs," as Caltech professor David Goodstein describes it in his book *Out of Gas*. The most popular leaders of this movement also envision a massive "die-off" of the world's population, along with the end of industrial civilization.

There is only so much oil worth pumping out of the ground. Peak-oil theory claims that once it's half gone, the rate of pumping will reach an all-time production peak and start to decline. The peak will herald the beginning of an "earth-shattering crisis," as one author puts it. The world economy and, most likely, the world's population will decline right in step with oil production. According to peak-oil theorists, the oil is about half gone. Our time is up.

Goodstein, a physicist, says that "until the 1950s, oil geologists [believed] that the same rate of increase [in oil production] could continue forever." And geologists say that economists think this still. But I can find no evidence that anyone has ever believed in limitless oil. Back in the 1800s, a famous economist named William Stanley Jevons predicted peak coal in England far too early. And patent-medicine salesmen, hocking "rock-oil" remedies, predicted peak oil just before Edwin L. Drake drilled the first oil well in Pennsylvania. (Before then, people got oil from natural oil seeps.)

Starting in 1979, the *Mad Max* film trilogy painted a bleak and violent picture of a world plagued by oil shortages that cause a nuclear war. Since then, predictions of a similarly grim economic future have become attached to peak-oil theory.

Peak-oil geology has fascinated me since 1998, when I read a *Scientific American* article by two leading peak-oil geologists. Pursuing the topic more recently, I found its basic tenets showing up in mainstream arguments over U.S. energy policy. One such policy—that the U.S. military is to achieve "energy independence" through subsidies for liquid fuels derived from coal—is backed by the Departments of Energy, Defense, and the Interior.

As with the idea that we will "wreck the economy," fear of peak oil is counterproductive. Peak-oil scare tactics aid in the push for liquid coal and synfuels. Using these can nearly double carbon dioxide emissions. Worse still, overblown claims of economic collapse have led, naturally enough, to the erroneous conclusion that peak oil will solve the climate-change problem. This makes it easier to accept the push for liquid coal.

Peak-Oil Theory

In 1956, oil geologist M. King Hubbert predicted that U.S. oil production would peak between 1965 and 1972. It peaked in 1970. He also predicted that world oil production would peak between 1995 and 2000. He did not, however, predict an earth-shattering economic crisis at the peak. Experts base their predictions of peak production on graphs of historical production rates and clever extrapolations. These techniques involve neither geology nor economics and are easy to understand. For example, just read geologist Kenneth S. Deffeyes's fascinating book *Beyond Oil*.

More recently, peak-oil enthusiasts have added the *Mad Max*-flavored economic collapse to Hubbert's sober theory of peak oil. The collapse is most clearly explained by electrical engineer Richard C. Duncan, one of the most popular peak-oil proponents on the Web. (In 2007, Google listed 450,000 Web pages referring to him.) He claims the "world population will decline to about 2 billion circa 2050." Since the world's population is currently over 6 billion, that would mean over 4 billion would die—over sixty times more than died in World War II.

C. J. Campbell, a petroleum geologist and the leading peak-oil expert, also believes world population will fall to "pre–Oil Age levels," which would imply even more deaths. Richard Heinberg, the most prolific peak-oil author, tells us this is not "necessarily such a bad thing" because it "just means a return to the normal pattern of human life—life that is in tribes or villages" (see "The

The Peak-Oil "Die-Off"

The World's population has grown in parallel with oil production to its present level of 6.4 billion. … It is hard to avoid the conclusion that this Century will see the population fall to close to pre–Oil Age levels.

—C. J. Campbell, leading peak-oil geologist

The recent fossil-fuel era has seen so much growth of population and consumption that there is an overwhelming likelihood of a crash of titanic proportions. … Verbal and mathematical logic, joined with empirical evidence, make an airtight case: we're headed toward a cliff.

—Richard Heinberg, most prolific peak-oil author

Perverse as the comment may seem, I don't think collapse, in this instance, would necessarily be such a bad thing. As Tainter points out, collapse really just means a return to the normal pattern of human life—life, that is, in tribes or villages. … Perhaps peak oil at last provides the word "sustainability" with teeth.

—Richard Heinberg

Peak-Oil 'Die-Off" for his full quote). But Heinberg, a new-age journalist, was predicting this die-off even before he latched onto peak-oil theory.

What Happens after the Peak?

Oil production will certainly peak, and perhaps it already has. But what about the worldwide economic collapse? Will that certainly follow? The world did experience a peak in oil production in 1979, when the Organization of Petroleum Exporting Countries (OPEC) cut production and raised prices. Production declined sharply for four years and did not surpass the 1979 peak again until 1989. This provides a real-world test of the peak crisis theory.

So what happened when world oil production suddenly stopped rising and started falling in 1979? The world did not shatter; instead, it kept growing. Moreover, it outdid OPEC, cutting oil use more than OPEC had intended to cut production. Deffeyes, the most respectable peak-oil geologist, says we're now sliding over and down the final oil production peak. But by his calculation, the decline in oil production for the first five years after the peak, the period he's worried about, will be considerably less steep than the decline after the 1979 peak.

Deffeyes is a Princeton geologist and, for my money, by far the most interesting of the peak-oil experts. He has nominated November 24, 2005—Thanksgiving of that year—as World Peak Oil Day. Better yet, in *Beyond Oil,* he gives his exact formula for the peak, which we will soon check. Figure 1

Figure 1. The 1979 OPEC Oil Peak Was Sharper than Deffeyes's Oil Peak

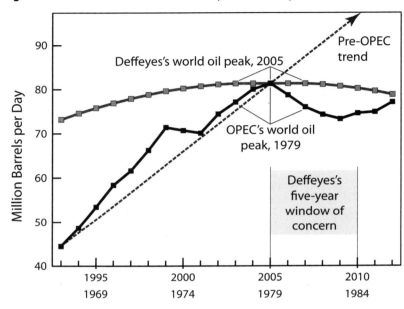

The graph aligns the 1979 peak in world oil production caused by OPEC with the world's final oil production peak, as predicted by Deffeyes, so that the two can be easily compared. Peak-oil theory predicts a smooth peak. Consequently, the shock to the world economy was much greater during the first six years of OPEC's peak than the economic shock expected from the current peak—if this is, in fact, the peak.

shows Deffeyes's predictions about world oil production. The peak in production centers on 2005, and the graph is based on his "logistic" formula and his value of a 10 percent drop-off by 2019. Deffeyes is optimistic that in fifteen years we will find adequate "renewable, non-polluting, sustainable" energy sources, but he says he's worried about the first five years, 2005 to 2010. "What can we expect on the five-year time scale? … Get acquainted with parsnips and rutabaga." In particular, he's worried that "war, famine, … and death … are serious possibilities." But in the first five years, production would drop only 1.4 percent. Why is he so worried?

He's concerned that world demand for oil was growing at almost 2 percent per year before World Peak Oil Day and that growth will have to stop. With Deffeyes's prediction of slightly negative growth in production, we would fall behind a full 10 percent in five years. That's a lot to be short of gasoline.

However, in 1979, the world's use of oil had been rocketing up more than twice as fast as in recent years. Five years after the 1979 peak, oil supply had fallen about 20 percent below its upward trend. So the shortfall after the

Figure 2. Peak Oil Had Little Effect on World GDP

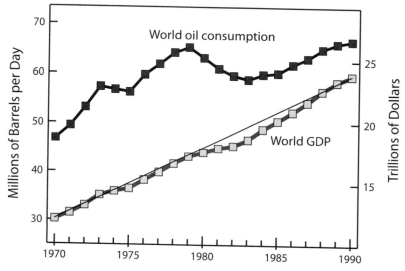

Although the 1979 peak in oil production and consumption was sharp, it did not have a catastrophic effect on world economic production. The final peak in oil production will not cause a global economic crisis killing billions, as predicted by a number of peak-oil proponents.

1979 peak was twice as severe as the shortfall Deffeyes foresees as likely to cause war, famine, and death. So what actually happened in the five years after the 1979 peak?

During that time, when total world oil production and consumption fell 8 percent, world gross domestic product (GDP) grew by 13 percent (see Figure 2). I'm not saying OPEC's impact was painless, but 13 percent growth in five years is not a calamity. The first few years were tough times—the poor suffered, and the rich were annoyed—but the world economy did not stop growing.

World oil production did not make it back to its 1979 peak until 1989, and in those ten years, world GDP grew 35 percent. Supply reductions tend to send prices soaring, and at first they did. But by 1986, with world supply down 8 percent from its peak, the price of oil was down 70 percent from its peak. How could a drop in supply cause prices to collapse?

Ahmed Zaki Yamani, the Saudi oil minister and a decent economist, foresaw this and tried to rein in OPEC's price increases in 1979. He succeeded a bit, but he knew it was not enough. Yamani knew high prices were a two-edged sword. They pried trillions of dollars from the purses of consuming nations. But what the peak-oil proponents deny—and what Yamani understood—was that consumers do not sit idly by and watch this happen. When OPEC's prices soared, consumers, including businesses, cut demand so much that they more

Figure 3. World Oil-Supply Predictions

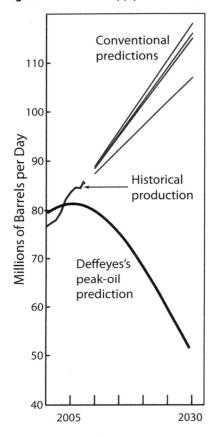

In his book *Beyond Oil*, Deffeyes predicts the oil production peak graphed here. So far, it has not happened. Actual world production is shown from 2000 through the first four months of 2008. These data are from the U.S. DOE (see endnotes for more details).

The National Petroleum Council, in its July 2007 report *Hard Truths*, cites four conventional predictions for oil production in 2030. From highest to lowest, these are from the U.S. Department of Energy (DOE); the International Energy Agency (IEA), a group of industry consultants; and an average of international oil companies' predictions.*

than made up for OPEC's supply cuts. Supply did not fall short of demand.

In Figure 1, notice the huge gap that opens up, in the 1980s, between the oil supply trend and the actual supply of oil. This gap is the result of conservation. Conservation punished OPEC's excesses for decades. Peak-oil geologists may know their oil. But Deffeyes confesses, "I emphatically do not understand economics." Someday the market will teach him the same lesson it taught OPEC.

Some might argue that the 1979 and 2005 peaks are fundamentally different. The 1979 peak was not the real peak, and world oil production surpassed 1979 levels ten years later. When the real peak comes, there will be no going back. But knowing that the peak is final will only cause markets to adjust to falling supply more quickly and decisively than they did in 1979.

Has the World's Oil Supply Peaked?

November 24, 2005, has come and gone. As Figure 3 shows, it wasn't exactly World Peak Oil Day. But if there's a sharp decline in oil production after 2009, Deffeyes will not have been so far off. The trouble is, we just don't know, and prognosticators have a long history of jumping the gun.

In 1919, for example, the director of the U.S. Bureau of Mines predicted that "within the next two to five years the oil fields of this country will reach their maximum production." In 1943, Secretary of the Interior Harold Ickes published an article referring to U.S. oil production with the title "We're Running Out of Oil!"

It can be a bit difficult for those of us who are not geologists to believe predictions of an imminent peak in oil production because such predictions are in sharp disagreement with the forecasts of the oil industry and government agencies.

If most experts believe the peak-oil proponents are wrong, why take them seriously at all? One reason is that the experts themselves have been wrong of late. Between 2005 and 2007, the DOE cut its prediction of the 2010 world oil supply by 4 percent. That's quite a lot for such a short-term prediction. Something is changing unexpectedly. Since 2005, in spite of prices that might have stimulated more production

Peak-Oil Economics Unscrambled

History shows that the world economy did not collapse when oil supply peaked sharply in 1979, so where have the peak-oil geologists gone wrong in their economic thinking? Peak oil's *Mad Max* economics assumes markets work like this:

The demand for oil increases as wealth and population increase. After the peak, the supply of oil will fall. Therefore, supply will not meet demand, and a crisis will destroy the world economy.

However, basic economics predicts that unless the government interferes, markets will work like this: The demand for oil increases as wealth and population increase. After the peak, the supply of oil will fall. Falling supply will cause the price to rise, and that will cause people to use less. Demand for oil will fall until it equals supply.

That is what happened in the early 1980s. Deffeyes, the Princeton geologist, knows both theories and explains in his book *Beyond Oil* why he thinks basic economics is wrong. "For the first time since the Industrial Revolution," he begins, "the geological supply of an essential resource will not meet the demand."

This could contain a grain of truth if Deffeyes's conjecture on government intervention is right. Markets have worked for all essential resources. But Deffeyes is worried that the law of supply and demand is about to break down for the first time in 250 years. Deffeyes remembers that: "Virtually all economists visualize it as price increases that bring supply and demand into a new equilibrium." Exactly. By equilibrium he just means supply equals demand. But after remembering the reason supply will equal demand, he rejects it.

"That outlook is widespread," Deffeyes says. "It must be something that Gerber puts in baby food." He doesn't believe that price will do the job. Instead, he has another theory, which he supports with two examples from history:

- ▶ "Historically, President Nixon regulated the oil price."
- ▶ "President Roosevelt had us carrying little red and blue gasoline ration coupons."

Deffeyes is right that, if the government intervenes, it can break the market and then demand will fail to equal supply. According to Deffeyes, this is why, after 250 years, the market for oil will break down when oil production peaks. This will cause peak-oil's economic collapse.

Deffeyes argues that the government will intervene because "when the situation gets serious, there will be immense political pressure to 'do something.'" But Deffeyes overlooks what happened after Richard Nixon regulated the price of oil. By the end of the OPEC crisis, virtually the entire elaborate system of oil price controls, gasoline price controls, and quantity rationing had been eliminated. This took immense political wrangling, but eventually there was widespread agreement. The country learned something back then, and I don't think it's about to forget it and cause the collapse of the American economy—or the world economy.

What If the Price of Oil Went to $200?

The market will keep demand equal to supply, as it always has—except under price controls. But if supply falls, the price of oil will increase to reduce demand. The whole question of peak oil boils down to one not-so-easy question: How high will prices go?

In August 2005, Matthew R. Simmons, chairman of an energy investment banking firm in Houston and author of a book about peak oil, bet $5,000 that oil would average $200 a barrel in 2010. A recent book on peak oil is entitled *The Coming Economic Collapse: How You Can Thrive When Oil Costs $200 a Barrel*. What would happen if this prediction is right?

In 1979, the price went to $90 (in 2007 dollars), but back then the world was using almost twice as much oil per dollar of income, so a $90 price then was almost as hard to take as a $200 price now. At $200 per barrel today, the United States would spend 11 percent of GDP on energy, instead of 10 percent as in 1980. But gasoline would still cost less in the United States than it does now in some European countries. If the oil price suddenly shot up to $200, it would cause a world recession, as it did before. But as before, economic growth would soon resume.

than expected, supply stayed flat through 2007. This fits with Deffeyes's flat peak. But in early 2008, oil supply is on the rise again.

Peak-oil geologists suggest that although the oil industry officially predicts the peak is years away, industry insiders actually know better. In fact, there is one good reason the industry might not want to admit the peak is near. The oil industry remembers what happened to it when the price collapsed in early 1986. The entire oil industry fell on hard times. Exxon's profits fell almost to zero. The world had partially kicked the oil habit. That's the last thing a pusher wants to see. And if the addicts knew oil production was about to peak and then decline forever, they might decide to look for a rehabilitation program sooner rather than later.

But perhaps the oil industry has a different angle. What if the industry could find a new but expensive supply of oil? Take, for example, the oil shale that Presidents Ford and Carter wanted to turn into synfuels. Converting coal to oil is another possibility. The United States leads the world in both these resources. "We could be the New Middle East." Does that sound far-fetched? The words belong to Theodore K. Barna, assistant deputy undersecretary of defense.

Liquid Coal: The Dark Side of Peak Oil

Peak-oil proponents deny that more oil will be forthcoming at higher prices. But one surefire path leads to more gasoline. Chemists discovered the dirtiest antidote for peak oil eighty years ago: the Fischer-Tropsch process, which turns coal into gasoline.

This is not a theory. This is what powered the German Luftwaffe during World War II, as well as much of South Africa when the United Nations adopted an oil embargo against that country in 1987. In 1938, Germany consumed 44 million barrels of "oil," of which 10 million was synfuels from coal. By 1943, the country's synfuels output had reached 36 million barrels per year. Think of it as Germany's response to peak oil—the country's own personal peak and decline, which the Allies caused by cutting off Germany's oil imports.

Of course, chemists and engineers have refined the process over the years. Today, Montana, whose governor has been pushing liquid coal in recent years, could produce gasoline for the equivalent of about $55 per barrel of oil. This has not yet happened because investors are afraid the price of oil will fall back below $55 as soon as they build a coal-to-gasoline plant. In addition, they might have to pay a global warming charge.

The last time oil was this expensive, the price did drop back to $20 a barrel for more than a decade, so the investors' fears are warranted. But if we truly start running out of oil, the price will never again drop below $55, at least not for long. Then investors will build synfuel plants, just as the Germans did seventy years ago.

Making gasoline is possible, but is there enough coal? Deffeyes assures us that "the world has at least a 300-year supply of coal." To his credit, Deffeyes admits that coal will come in as oil runs out. "I hate to say it," he writes, "but we likely will be forced to choose either increased pollution from coal or doing without a significant portion of our present-day energy supply." There can be no doubt that between these two, people would choose "increased pollution from coal." That is why scaring people about running out of oil could be disastrous for climate change.

The Air Force Wants Liquid Coal

On October 22, 2007, an Air Force C-17 Globemaster III took off—not usually a newsworthy event. But the Globemaster is the biggest user of jet fuel in the military, and it was using a 50 percent blend of Fischer-Tropsch fuel.

The flight was a test of whether liquid coal might work for the U.S. Air Force as it worked for Germany's air force. It did. The test flight was part of the Assured Fuels Initiative, which the Department of Defense set in motion in 2001. One of the initiative's objectives is for the Air Force to get half its fuel from domestic sources by 2016. It would take 110,000 barrels per day of domestic fuel to reach this target level. Since the United States currently pumps about seventy times this much domestic oil per day, it would be easy to accomplish this goal without a synfuels program. So what is the real reason for this initiative?

The initiative appears to provide one legitimate benefit: Synthetic fuel should work better across a wider variety of vehicles than standard gasoline, diesel, or jet fuel. But this is unlikely to be an argument that wins support for major new subsidies to Big Oil. Of course, the military might also have in mind reducing the need to defend Middle East oil supplies, but replacing less than 2 percent of imports does not amount to much.

This leaves the fear of a peak-oil crisis as the major lever for gaining public support. And indeed, proponents of this initiative have not hesitated to play that card. "World petroleum supply trends indicate that the days of inexpensive oil may be over," says a report of the Task Force on Strategic Unconventional Fuels. The Energy Policy Act of 2005 mandated the formation of this task force, and the secretaries of Energy, Defense, and the Interior implemented it. The report continues, "Peaking global production … is already causing competition for supplies." An accompanying graph echoes those in the peak-oil literature and seems to show "world remaining oil reserves" shrinking to zero in about 2030.

The U.S. synfuels industry, which is just another name for the oil industry—Exxon, Shell, and the like, wants to regain its old subsidies and more. A military rationale, combined with a peak-oil scare, could be just the ticket. The oil companies are beginning to see peak oil as less of a threat and more of an opportunity. If the public is frightened into conserving, it's a disaster for Big Oil. If the public is frightened into subsidizing synfuels again, it could be a bonanza for the oil companies. Although most peak-oil proponents favor sustainable approaches in theory, they tend to dismiss conservation and alternative fuels as too little too late. All that remains is coal, shale oil, tar sands, and the fear of gasoline shortages. Big Oil owes the peak-oil proponents a great big thank-you, not for predicting the peak, but for helping frighten the public into subsidizing a move into synfuels and unconventional oil.

Peak Oil and the Southern States Energy Board

The Southern States Energy Board (SSEB), originally the Southern Interstate Nuclear Board, is an interstate compact that Congress approved in 1962. Sixteen states and two territories have also approved it. A July 2006 SSEB report recommends a list of subsidies for Big Oil, similar to those recommended by the federal Task Force on Strategic Unconventional Fuels. Here's how the SSEB justifies the subsidies: "America now faces a crisis of historic proportion: a liquid transportation fuels crisis. This Study shows that immediate implementation of 'crash' programs to ramp up production of domestic alternative liquid transportation fuels is the only way to insure against peak oil."

~

Sooner or later, the production of cheap conventional oil will peak. What should we do? We must not let predictions of doom either paralyze us or prompt us to take rash action. Rather, we should simply do what we should be doing anyway: taking measures to prevent climate change and end oil addiction. That will be sufficient. Conservation will work the most quickly and robustly in the short run, but we will need new technology for the long run. I sketch out a clear plan of action, aimed at both conservation and new technology, in Chapter 7.

If we fail to take such action, the market will do the job for us. But it will cost us far more. The world may see an extra recession or two, OPEC and the oil companies will make a killing, and the world will burn far too much fossil fuel while converting coal, tar sands, and shale oil into gasoline. Climate change will accelerate.

Should we choose to subsidize Big Oil, this will worsen our addiction at taxpayer expense, just as it always has. Our oil companies will make a slightly larger killing, and OPEC will make a bit less. Climate change will be even worse than if we let the market work alone, and the world price of oil will be just a bit lower due to the extra supply. The Chinese, who are very short of oil, will thank us.

chapter 4

Is the Globe Warming?

I don't want to wait around until the house burns down till I decide whether it's a serious fire or not.

—Oilman T. Boone Pickens on climate change, 2008

TWO MYTHS HAVE CLOUDED our understanding of climate science. Believe the first—that climate science is still too uncertain to serve as a guide for action— and we will do nothing. Believe the second—that the signs of imminent disaster are so obvious that we no longer need science—and we may waste trillions.

Fortunately, an easy solution is at our disposal: Believe the Intergovernmental Panel on Climate Change (IPCC), and believe this chapter's quote from T. Boone Pickens. They both make sense, and together they provide the clarity we need. The IPCC is the world's leading scientific authority on global warming, and T. Boone Pickens is a hard-nosed oil billionaire.

Science is cautious. It does not accept the result of one experiment or test but demands cross-checking by many scientists. Consequently, science is slow to reach a firm conclusion, and scientists are prone to say, "It's probably like so, but we aren't sure yet." And that is exactly why we should believe them. Don't trust those who jump to conclusions or have an ax to grind; they are the mythmakers.

The IPCC tells us that human activity is probably causing most of the global warming but that the IPCC isn't sure about that yet. They're scientists. They are only 90 percent sure. That leaves the door open for the first myth—that we don't know enough to do anything yet. That's where T. Boone Pickens comes into the picture.[1] He admits the scientific uncertainty but draws the obvious conclusion: If our house is on fire, we should not wait for the scientists to tell us precisely how serious it is before we do something about it. The scientists won't be completely sure till it's too late.

In this chapter, I first investigate the sources of the two myths. Then I take a closer look at just what the IPCC has to say and why it makes sense to get moving as soon as possible—which will be none too soon, given the sluggishness of international organizations.

Doubt and Uncertainty Is Their Strategy

A leaked memo reveals the origins of the first myth—that scientific uncertainty means we should do nothing about global warming. It was an internal memo of the Global Climate Coalition, an organization of major corporations that, from 1989 to 2002, fought attempts to reduce greenhouse gas emissions. In the 1998 memo, the group clarified its definition of victory: "Unless 'climate change' becomes a non-issue, meaning that the Kyoto proposal is defeated and there are no further initiatives to thwart the threat of climate change, there may be no moment when we can declare victory."

To the oil, coal, and auto companies that formed this coalition, victory was the defeat of the Kyoto Protocol and the end of all "further initiatives to thwart the threat of climate change." Those companies did not wait for scientific proof that their profits were threatened before forming their coalition just a few months after the United Nations organized the IPCC.

Wary of the new scientific initiative, the coalition focused on casting doubt on the science. The 1998 memo shows them chagrined to find they have been losing the battle, but it points to an opportunity: "The science underpinning global climate change theory has not been challenged effectively in the media." The memo also emphasizes the need to get "average citizens to 'understand' (recognize) uncertainties in climate science."

But as climate science turned up more and more evidence against the coalition's position, the group began to disperse. DuPont, British Petroleum, Shell, Ford, DaimlerChrysler, General Motors, and Texaco all left by 2000. Exxon stuck with the coalition until it became inactive in 2002. By that time, Exxon had found champions in the new Bush administration.

Among top Republicans, Frank Luntz may be the most renowned public relations specialist. He was the principal author of and pollster for Newt

---

1. Pickens's insight is supported by a difficult but brilliant paper by Martin L. Weitzman, a Harvard professor, "The Stern Review of the Economics of Climate Change."

Gingrich's "Contract with America." In 2002, Luntz advised the Republicans on techniques for "winning the global warming debate":

> The scientific debate is closing [against us] but not yet closed. There is still a window of opportunity to challenge the science. …
>
> Voters believe that there is no consensus about global warming within the scientific community. Should the public come to believe that the scientific issues are settled, their views about global warming will change accordingly. Therefore, you need to continue to make the lack of scientific certainty a primary issue in the debate. …
>
> Emphasize the importance of "acting only with all the facts in hand [bracketed note in original]. (*Winning the Global Warming Debate*, 2002)

Luntz warned that winning would not be easy, because the scientific debate was "closing against" the Republicans. So he urged them to "make the lack of scientific certainty a primary issue." That was an ideal approach for him to choose, because it takes science decades to nail down all the details in a complex field. Emphasizing "the importance of 'acting only with all the facts in hand'" completes the link between "lack of scientific certainty" and taking no action.

Of course, it doesn't really make sense to wait until "all the facts [are] in hand." We normally make intelligent decisions without scientific certainty. Someone puts one bullet in a six-shooter, spins the cylinder, and points the gun at your head. Don't worry; no action is needed. Science has not yet proved you will die. And it never will. Science will always put the odds of your being shot at one in six—an uncertain outcome. Luntz would have us act "only with all the facts in hand"—that is, right after the trigger is pulled.

Global warming is not as dangerous as a gun to your head, but as with the gun a real chance of catastrophe exists. Ignoring such risk because of a "lack of scientific certainty" is not a sensible strategy.

The argument Luntz pitched to the Republicans is psychologically powerful, though not new. The tobacco companies used the same strategy for years to cast doubt on the science about cancer and cigarettes. The idea that the smallest scientific uncertainty indicates that we should do nothing is a recycled myth that goes under the code name of "sound science."

"Sound Science"—a Short History

While I had heard people draw parallels between the denial of cancer risk by the cigarette industry and the denial of global warming by the oil industry, I was surprised to learn that an organizational and strategic link exists as well.

"Sound science," as George Orwell might have predicted, is the code name for questioning the mainstream science behind the hazards of cigarette smoke, global warming, and other phenomena.

In 1993, Philip Morris hired a public relations firm to secretly set up the Advancement of Sound Science Coalition. Its goal was to convince the public that secondhand smoke was not a problem. By then, ten years of scientific studies indicated that secondhand smoke could be lethal. The result was a grassroots movement advocating no-smoking areas. Philip Morris was worried, because, unlike smokers, people exposed to secondhand smoke cannot easily be blamed for inhaling cigarette smoke. Legally, secondhand smoke was hazardous to the health of Philip Morris.

As it turned out, the science continued to point ever more strongly toward such health risks. Today, even Philip Morris admits on its Web site that "particular care should be exercised where children are concerned, and adults should avoid smoking around them."

However, in 1993, when Philip Morris launched the Sound Science Coalition, Steven Milloy—now the Fox News commentator on global warming—was a registered lobbyist working for a company that was receiving $40,000 a month from Philip Morris. And Milloy was calling the EPA's then recent study of secondhand smoke a "joke." That study reached milder conclusions about the danger of secondhand smoke than those now endorsed by Philip Morris itself.

By 1997, Milloy was executive director of the Sound Science Coalition. But in 1998, the press discovered the coalition was actually a front group for the tobacco industry. Once this was public knowledge, the coalition lost its value as a means of deception, and Milloy closed it down. But that same year, he opened the Advancement of Sound Science Center (at the same address as the Advancement of Sound Science Coalition) and used it to begin attacking global warming science. By 2000, Exxon was funding Milloy.

Scientists do not commonly use the phrase *sound science*. A search of the *New York Times* finds it used in only one story in the 1970s, and its new political meaning shows up only in 1986. The *New York Times* first reports its use in high-level politics in 1992, when President George H. W. Bush used it to attack the Food and Drug Administration.

It was also in 1992 that Philip Morris budgeted $880,000 to launch the Sound Science Coalition, kicking the term deep into Republican territory. Let's check back with political strategist Luntz as he teaches Republicans how to cast doubt on the science of global warming. Just before he warns that "the scientific debate is closing against us," he says, "The most important principle in any discussion of global warming is your commitment to sound science."

Evidently, Luntz's Republican students took their lessons seriously. Compared with only sixteen mentions in the *New York Times* between 1970 and 1992, sound science shows up in 143 *New York Times* stories since then. A Google search for the term on the official White House Web site found it on 314 pages.

Steven Milloy spent years pressing the tobacco industry's claims concerning secondhand smoke. But the scientific debate closed against Big Tobacco, and Philip Morris and R.J. Reynolds now admit they had it exactly backward. What they ridiculed as "junk science" was actually sound, mainstream science.

Milloy has now spent years pressing the oil industry's claim that carbon dioxide does not contribute to global warming. In November 2007, on Fox News, Milloy was busy as usual attacking scientists. Commenting on a United Nations report on global warming, he said:

> This glib statement overlooks the fact that from 1940 to 1975 globally-averaged temperature declined. ... If there's a cause-and-effect relationship between CO_2 and temperature in the last 50 years at all, it seems to be slightly in the opposite direction from what the U.N. claims.

But the statement Milloy calls "glib" is the central conclusion of a four-volume, 2,000-page United Nations report summarizing five years of research by thousands of scientists and endorsed by roughly a hundred countries. The temperature decline that Milloy refers to as "overlooked" is in fact an aspect of global warming that scientists have studied extensively. The discovery that sulfur emissions caused the decline is a key part of the evidence that CO_2 emissions do cause global temperatures to rise.

Milloy's second Advancement of Sound Science group has folded as the scientific debate has all but closed against him again. The battle is not over, but Big Oil is forced now to shift tactics and become more discreet.

What Does Exxon Really Want?

As a business article in the *New York Times* put it recently, Exxon is "unapologetically geared toward generating returns [profits] for its shareholders." Of course, all corporations are focused on profits, and that's why economists can sometimes predict what they will do. So what does economics predict about Exxon's global warming strategy?

Because Exxon's profits go up and down with the price of oil, the company wants high oil prices. That's a snap. But those prices are hard to control, even for Exxon. Only two influences are powerful enough to make much difference: OPEC and the Kyoto Protocol.

OPEC pushes oil prices up by restricting supply. Kyoto pushes prices down (a little) by restricting demand. Of course, that's not the point of the Kyoto Protocol, but that's one thing it does, and that hurts oil company profits.

So economics—and common sense—make a clear prediction: Exxon wants OPEC to succeed and global warming policies to fail.

Of course, since Exxon wants to maximize its profits, it's unlikely the company would ever admit to all that. It would make the company even more unpopular than it already is, which hurts business.

As attitudes shift in favor of global warming initiatives, Exxon's job becomes more difficult. To be taken seriously, Exxon must now appear to take global warming seriously—and it does appear to. Exxon wants in on the public discussions—wants to be "at the table." As Charles Territo of the Alliance of Automobile Manufacturers explains, "If you're not at the table, you're on the menu." And as Kenneth P. Cohen, Exxon's head of public affairs, told reporters in June 2007, "We're very much not a denier, very much at the table with our sleeves rolled up." But on the sly, Exxon still fights to discredit global warming.

Holly Fretwell's new book is for children. It's called *The Sky's Not Falling! Why It's OK to Chill about Global Warming*. Fretwell, an economist, claims her "expertise is not in climate science," yet after a short discussion, geared for sixth-graders, of what she claims are climate science fallacies, she concludes, "This all makes it highly unlikely that the current warming trends are a result of human activity."

In December 2007, when Fretwell was asked about the group that funded her book, she replied that her organization "does accept a small amount of money from Exxon to help cover our general overhead expenses. I can only assume that this support comes because they like what we do."

Of Islands and Sea Levels

Exxon is worth about half a trillion dollars. Ross Gelbspan, a Pulitzer Prize–winning journalist, rather less. But he enjoys taking on the giant. Al Gore, for one, has commended him for his efforts, and he deserves the praise.

But page 2 of Gelbspan's 2004 book *Boiling Point* begins with a curious statement: "The evidence [for global warming] is not subtle." Gelbspan finds the case for global warming terrifyingly obvious. But if the evidence really is so obvious, why don't the scientists notice? Why do they keep doing all these complicated studies and end up only 90 percent sure? Are they a bit dense? Perhaps they should read Gelbspan's book.

Gelbspan's certainty that global warming is obvious runs through his work as a reporter, making him incautious. Consider this excerpt from *Boiling Point* about a group of Pacific islands:

> In November 2000, officials began the permanent evacuation of more than 40,000 people from their traditional home. As the British newspaper *The Independent* noted, "[this] could be the dress rehearsal for millions of people around the globe affected by rising

sea levels." ... The islands are just 12 feet above sea level, and water levels are rising at 11.8 inches per year.

Gelbspan tells us—based on an article in *The Independent*—that the sea level is rising 11.8 inches per year due to global warming. But an experienced reporter writing his second book on global warming should have noticed something fishy about 11.8 inches per year. That really is awfully fast.

So how might an investigative reporter proceed? First, a close reading of the source newspaper article, which can be found on Gelbspan's Web site, reveals it does not say the sea level was rising 11.8 inches per year. Instead it says "The islands ... are sinking 11.8 inches a year." That's a little different.

To check further, a reporter might next try the IPCC's 2001 report. Download the *Summary for Policymakers* from the group's Web site, and search for "sea level." The second hit reads, "Global mean sea level: Increased at an average annual rate of 1 to 2 mm during the 20th century." That's in Table 1. There are about 25 millimeters to an inch. Two millimeters annually is less than a tenth of an inch per year.

So 11.8 inches per year is about 100 times too fast to be caused by global warming. The islands' problem is not the tenth-of-an-inch per year rise in sea level. The problem really is that the islands are sinking. Here's a news report from 2000 explaining why.

> The move from the Duke of York group [of islands] is mostly due to a spectacular clashing of tectonic plates. The shift is extremely violent and this month saw a magnitude eight earthquake and several in the seven range. ... The islands are sinking 30 centimetres (11.8 inches) a year. (Michael Field, *Agence France Presse*, November 28, 2000)

The problem really is that the islands are sinking, and they are sinking because of plate tectonics—that is, one part of the earth's crust is sliding under another. This has nothing to do with global warming.*

Unfortunately, Gelbspan's misstatement of the facts appears to be part of a pattern in which Gelbspan and some other members of the press inadvertently undermine the credibility of the science of global warming by overstating its conclusions. For example, in the same book, Gelbspan says, "Were the Greenland Ice Sheet (or a substantial part of the West Antarctic Ice Sheet) to slide into the oceans, it could cause a rapid rise in sea levels. Since about half the world's population lives near coastlines, the consequences could be chaotic."

"Slide," "rapid," "chaotic." All possibly true on the centuries-long time-scales that climate scientists normally consider. But when I read that passage, I formed an image like one in an old-time newsreel, in which someone breaks a bottle of champagne across a ship's bow, and the ship slides into the water with a great splash. What Gelbspan and other reporters need to point

Global Warming by the Numbers

Three numbers are key to a basic understanding of global warming.

Temperature has increased:
  1° Fahrenheit since 1950

CO_2 has increased:
  about 1/3 since 1750

Sea level is rising:
  about 1/10 inch per year

If nothing is done about global warming, in the future these trends will likely accelerate.

out when they say "rapid" is that in a worst-case scenario—beyond anything the IPCC predicts—"rapid" means Greenland's ice will take 100 years to slide into the sea and the sea level will rise about half an inch per year.

Warning of extreme possibilities is valuable so that people can consider the risks. But reporting extremes as if they are the likely outcome, and reporting them in misleading language, ends up making people more skeptical of the science—to the delight, I am sure, of the oil companies.

The Scientific Consensus

Some reporters have let us down, as have a few scientists, some in the pay of Exxon. But the vast majority of scientists are true to scientific principles, and they are speaking to us clearly. The IPCC does a remarkable job of reflecting the scientific consensus, and it deserves our attention.

The IPCC's 2007 climate-change report gives us the scientific answer to the central question of climate change: Is human activity responsible for global warming? But to understand the answer, you must think like a gambler. If you ask a gambler: "Will next year be the hottest on record?" he will refuse to say yes or no. Neither will scientists. They will give you the odds. Scientists have reached a solid scientific consensus, and it tells us what we need to know. Here's how the IPCC puts it in its 2007 report:

> Most of the observed increase in globally-averaged temperatures since the mid-20th century is very likely due to the observed increase in anthropogenic GHG concentrations.

Here's the IPCC's conclusion in plain English:

> The odds are at least nine in ten that over half of the increase in global temperature since 1950 is due to human activity.

Nine in ten means a 90 percent chance, and that is how the IPCC defines the phrase "very likely." To avoid sounding too geeky, the researchers have redefined certain English phrases to refer to specific probabilities. "Very likely" is one of them. Of course, when a glass is 90 percent full, it is 10 percent empty, so it's also true that there's a one in ten chance that nature—not humans—caused most (not all) of the global warming since 1950.

Because the IPCC does not make these statements unless all the roughly 100 IPCC nations agree, the statement must be weak enough to get the most skeptical nation to consent to it. At present, scientists have produced no other consensus statement, so for policy purposes it seems best to rely on the IPCC.

Why Act Now?

If scientists are not yet completely sure what's causing global warming, why not wait for them to figure it out? Actually, two good reasons make it urgent that we act now. First, science is already 100 percent sure the world faces a serious risk. Second, the world is extremely slow to organize.

We Are at Risk. Science is not sure that unchecked global warming will cause a catastrophe. But consider the question from the other perspective. Science is not sure the world is safe—in fact, it's not even 10 percent sure. The scientists in the IPCC readily admit their small uncertainty, but the deniers never admit their much greater uncertainty. That is the difference between science and propaganda.

In spite of uncertainties, the IPCC's conclusions tell us that the scientists are 100 percent sure the world is at risk and that the risk is not small. When we know about a risk—say, of a house fire, a car accident, or a terrorist attack—we take precautions to lower that risk. The question is not "Should we do something?" but "How much should we do?"

The IPCC's cautious scientists don't tell us how much to do; they only describe possible changes in temperature and what some of the side effects might be. They present six "equally sound" scenarios based on expected global temperature increases in the twenty-first century. The estimated increases from the 1990s to the 2090s range from 3 degrees Fahrenheit in the most optimistic scenario to 7 degrees Fahrenheit in the most pessimistic.

The gray bar on the right in Figure 1 indicates the uncertainty about the predicted temperature for one of the IPCC's scenarios. Assuming the six scenarios are equally likely, there is a 5 percent chance that the temperature will increase by more than 9 degrees Fahrenheit by 2095.

A temperature increase of 9 degrees brought us to the present balmy conditions on Earth from the depths of the last ice age, when glaciers extended from the North Pole halfway down Long Island. Another 9-degree rise would cause changes of a similar magnitude. Citizens of Washington, D.C., might be building dikes, and temperatures there would top 100 degrees thirty days out of the year instead of just one. In the next century, things would almost certainly get worse.

The Trouble with "Obvious"

If we convince people that they can prove that global warming is serious just by noticing hot weather and glaciers melting, they will think they do not need the help of scientific investigations.

Then when the weather turns cold for a few years and some glaciers stop melting, they will feel disillusioned.

The climate changes because of both human activity and natural forces. Inevitably, the natural forces will play some tricks. That is why we need the scientists—to sort out what is natural and what is not and provide a clear, steady answer.

The IPCC is actually noncommittal, refusing to give the odds on their scenarios. But to be fair, if we again assume the scenarios are equally likely, we find there is also a 5 percent chance that, even with no effective climate policy, global temperature will rise only 2.5 degrees Fahrenheit by 2095. The consequences would be milder.

We can hold our breath and hope for the low number. We have one chance in twenty. But there's just as good a chance of drawing the unlucky 9-degree warming.

When it comes to serious dangers, a 5 percent chance is high—ten times greater than the chance you'll have some type of house fire in the next year. But few people go without fire insurance. Guarding against such risks is clearly worthwhile, and it's not such a good idea to wait until the house is on fire before buying insurance. The risk of a fire is reason enough.

The World Is Slow to Organize. The second reason to act now is that we are slow. Fixing the climate requires that we take two steps, organizing and acting. Organizing is slow but cheap. There is simply no excuse for not getting organized as soon as possible. To organize quickly, we should postpone the squabble over how strict the policy should be. We can start out with a policy that is not too expensive but that's easy to adjust once it's in place.

This approach is the opposite of what happened with the Kyoto Protocol. In Kyoto, most of the effort went into arguing about how strict the caps would be. But because China, India, Brazil, Australia, and the United States were unhappy with the caps, they rejected the policy itself. Third world countries signed on, but only after getting full exemptions. Fifteen years later, we are still trying to agree on an organizational framework. This is the slowest path to getting organized. As a result, in 2007, the world emitted CO_2 25 percent faster than it did in 1992, when the United Nations started the process

~

Scientists are uncertain about the impact of human activity on the climate, but they are sure we are running a huge risk on our present course. As with the risk of fire, accident, or terrorist attack, a grave risk requires action. No global warming denier would suggest waiting to be sure of a terrorist attack before taking precautions.

Both the magnitude of the risks and the world's slowness to organize call for a crash program to construct a sound and effective international organization. Progress will be most rapid if we agree on a structure we can start with before adjusting the policy to full strength. By the time the organization is in place, the science will be clearer. This should make it easier to agree on the tough policies that we will likely need.

Figure 1. Two of the IPCC's Six "Equally Sound" Scenarios for Global Warming Analysis

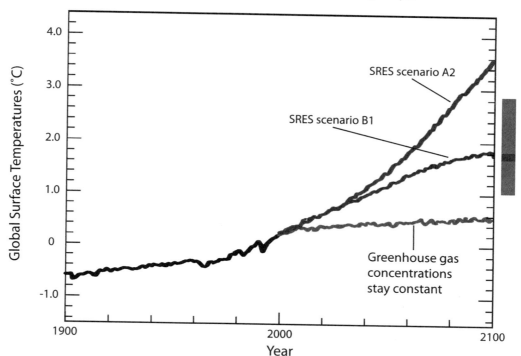

In 2008, atmospheric CO_2 reached a concentration of 386 parts per million (ppm), up from 280 ppm in 1750. The concentration of all greenhouse gases in 2008 is 485 ppm on a CO_2-equivalent basis. The three scenarios above (constant greenhouse gases, B1, and A2) show what might happen by 2100 if greenhouse gas concentrations increase not at all, to 600 ppm, or to 1250 ppm (and CO_2 increases correspondingly not at all, to 490 ppm, or to 850 ppm). The corresponding temperature increases by 2100 are estimated to be 0.6, 1.8, and 3.6 degrees centigrade (or 1.1, 3.2, and 6.5 degrees Fahrenheit).

The IPCC's "Special Report on Emissions Scenarios," (SRES) published in 2000, describes six marker scenarios which do not include additional climate policies beyond the ones in effect in 2000. Scenario B1 is the most optimistic of the six, and A2 is the second most pessimistic . The IPCC estimates that there is a 90 percent chance that the B1 scenario will result in temperatures in the year 2100 that fall within the range of the bar to the right of the graph. Source: Figure 3.2 of the IPCC's "Climate Change 2007: Synthesis Report."

chapter 5

Cheaper than Free?

Climate protection would actually reduce costs, not raise them ... because saving fossil fuel is a lot cheaper than buying it.

—Amory Lovins, *Scientific American*, 2005

IF PEAK-OIL PROPONENTS are the pessimists of the energy world, physicists are the optimists. Peak-oil buffs believe that having less oil will "end civilization as we know it," while energy guru Amory Lovins tells us that "oil problems will fade away" and that "displacing most, probably all, of our oil ... makes money." Lovins thinks that oil production will peak, not because we'll run out, but because we'll realize it's a waste of money and largely stop using it.

In the early days of the first OPEC crisis, when the Organization of Petroleum Exporting Countries (OPEC) tripled the price of oil, a number of physicists vigorously advocated conservation as the primary defense against OPEC. They claimed it was cheaper than increasing the supply of oil and sometimes cheaper than free. For example, insulation might save more in fuel costs than it costs to insulate. A couple of years into the crisis, in 1976, Lovins published, in *Foreign Affairs* magazine, a manifesto for the conservation movement. In "The Road Not Taken," he advocated a "soft energy path" to reverse the growth in U.S. energy use by conservation measures that would be cheaper than free. In spite of lacking a degree in physics, this made him perhaps the best-known member of what I will call the physics camp.

While many policy analysts and politicians, including Presidents Gerald Ford and Jimmy Carter, believed in stimulating conservation by raising energy prices, few believed this could be the primary solution to our energy problems. But as it turned out, it was mainly what put an end to OPEC's reign in 1986.

Without question, the physicists were right about conservation's importance. And they were right that, as Lovins puts it, conservation does not have to mean "discomfort or privation (doing less, worse or without)." Most of the physics camp, and many economists, agree that some conservation measures are cheaper than free. But Lovins goes further and claims that everything we need in the way of energy policy is cheaper than free. Is he right about this?

How Cheap Is Electricity Conservation?

As with peak oil, we can look to history to evaluate claims that conservation will be cheaper than free. Lovins's 1990 paper "Four Revolutions in Electric Efficiency" provides a historical test of this idea. It concludes that four electricity revolutions were in full swing with no roadblocks in sight (see "Electric Revolutions"). In short, he predicted that by now we could be using almost no electricity—only about 3 percent of what we used in 1990—and that this conservation effort could save us, counting all costs, over $200 billion a year. To be fair, he did not think we would take full advantage of these opportunities.

Lovins's starting point is that already in 1990, "the best technologies now on the market could save about 92 percent of U.S. lighting energy." However, for all electrical uses combined, he claimed that only three-quarters of the electricity used was unnecessary at that time. Moreover, Lovins tells us that his conservation measures would have cost eleven times less than the value of the saved electricity.

Next, he claims that the cheaper-than-free opportunities had doubled in the previous five years and would do so again in the next five and that he saw "no signs of this slowing down." Better yet, the cost of conserving would be decreased by three times every five years. (See "Predicting Conservation" for his calculations.)

As it turned out, between 1990 and 2005, electricity use went up 34 percent, not down 97 percent. It's hard to say exactly what went wrong, because Lovins doesn't leave behind documentation that others can check. But the point to remember is that counting on energy savings to happen on its own, even when the potential seems gargantuan and the monetary savings enormous, is risky business.

Hypercars and Formula One Race Cars

After predicting revolutions in electricity conservation, Lovins refocused on "Hypercars," vehicles designed to get such good mileage that they will, according to Lovins, "ultimately save as much oil as OPEC now sells." They were in the news on and off for ten years, so you may have heard of them, but do you know whatever happened to them? They sounded great. Were they too expensive? Were they underpowered? Let's follow their development to find out.

The story begins in 1981 with the McLaren MP4/1 Formula One race car—the first built on a carbon-fiber chassis. Carbon fiber is almost pure carbon. It is stronger than steel but much lighter. It is also much more expensive. In Formula One racing, where money is no object, carbon-fiber frames immediately rocketed in popularity. Ferrari has also used carbon fiber in a $500,000 supercar, as has Tesla Motors in its electric sports car, which has a base price of only $98,000.

In December 1990, while Lovins was writing about top-of-the-line cars that would get 60 mpg, General Motors was planning its Ultralite, a four-passenger, carbon-fiber car that could go 135 miles per hour. In April 1991, the company began chassis fabrication. By then, Lovins was reviewing state-of-the-art industry car designs. In July, when Lovins presented his ideas on fuel efficiency to a committee of the National Academy of Sciences, which was working on a report on fuel efficiency. Someone from General Motors heard the talk and invited Lovins to a sneak preview of the Ultralite. By December 1991, the company was showing the car to the press. Although General Motors claimed the car got 100 mpg at 50 miles per hour, the Environmental Protection Agency tested it at only 88 mpg.

By March 1994, Lovins was speeding toward the Hypercar:

> We are currently working with approximately 20 capable entities eager to bring Supercars [the original name for Hypercars] to market, and there are more entities joining the list almost weekly. Several are automakers. … There's been an astonishing flurry of licensing and other partnering arrangements just in the last few months with many of the key enabling technologies.

Lovins was guessing that he would see "significant production volumes starting around 1998 or 1999." He expected that by 2000 the end of steel cars would be in sight and that by 2005 "most, if not all, of the cars in the showroom will be electrically propelled."

Later that year, specifics of the car emerged. "Analysts at Rocky Mountain Institute have simulated 300–400-mpg four-seaters with widely available

Electric Revolutions

In a 1990 article, Amory Lovins predicted that four "revolutions" in electric efficiency would greatly accelerate the conservation of electricity. The revolutions were:

Technical progress. For example, better lightbulbs.

Markets for "negawatts." Negawatts are watts of electricity not used.

Cultural change inside utilities. For example, learning that conservation is profitable.

Reforms in regulatory philosophy and practice. In particular, "decoupling" profits from increased sales.

Predicting Conservation:
By 2005 We Could Use Only 3 Percent as Much Electricity

Lovins is famous for his command of facts and numbers, which seem to prove that amazingly cheap conservation is possible. But a close look at his numbers in "Four Revolutions" reveals that the individual claims not only sound amazing, but when taken together are completely unbelievable, as confirmed by history. (Warning: math ahead.)

First claim: "The levelized cost of that quadrupled end-use efficiency averages about 0.6 cents/kWh [per kilowatt-hour]."

"That quadrupled end-use efficiency" refers to the entire electricity sector, which Lovins says could have used four times less electricity in 1990. In 1990, the cost of electricity was 6.6 cents/kWh, so saving electricity for only 0.6 cents/kWh is eleven times cheaper than buying it.

Second claim: "We now can save approximately twice as much electricity as we could five years ago, but at only a third of the real cost. That is about a six-fold gain in cost-effective potential in five years, and nearly a 30-fold gain during the past 10 years. I see no signs of this slowing down."

Here Lovins tells us how fast things are getting better. Every five years we can save "twice as much" electricity as before, and he sees "no signs of this slowing." So his original "quadrupled" efficiency, a four-fold gain in 1990, doubles to become an eight-fold gain in 1995, then doubles to a sixteen-fold gain in 2000, and finally becomes a thirty-two-fold efficiency gain in 2005.

This would mean using thirty-two times less electricity—only 3 percent as much as in 1990. Meanwhile, with the cost of efficiency starting so low and dropping so quickly, the efficiency measures would cost almost nothing. Instead of conserving, fools that we are, we are using 34 percent more electricity and spending $290 billion dollars per year on it.

technology." To Lovins, this was not such a stretch, considering that he thought cars could get "more than 600 mpg with the best ideas now in the lab." Lovins's new concept that supposedly made all this possible was the idea of combining a car body like that of the Ultralite carbon-fiber body with an electric hybrid motor. Neither idea was new, but after combining them Lovins believed he had found a "powerful synergy between ultralight construction and hybrid-electric drive; the 1-plus-2-equals-10 equation." All this sounds impressive, but the theory does not quite add up (see "The Hypercar Fallacy").

"By spring 1996," Lovins says, "commitments to ultralight-hybrid development totaled ~$1 billion, recently doubling in less than a year." In early 1998,

Lovins urged the plastics industry to build one Hypercar for demonstration purposes, estimating it "could cost on the order of $10–100 million."*

Hydrogen Hypercars

By early 1999, with the rising interest in hydrogen, Lovins saw another opportunity for increased efficiency and cost savings. He would replace the Hypercar's hybrid motor with a hydrogen-fuel-cell motor. But this created a new hurdle— how to develop a hydrogen economy to support hydrogen-fuel-cell Hypercars. Lovins recognized that two problems, each insurmountable on its own, could be combined, using the logic of the one-plus-two-equals-ten equation. The combination would yield an efficient and even profitable solution.

In April 1999, he published "A Strategy for the Hydrogen Transition." It explained how, when Hypercars were parked at work, their hydrogen fuel cells could generate electricity and pure water for the buildings they were near. This would soon make hydrogen Hypercars the "dominant paradigm of the emerging hydrogen industry."

Even earlier, in 1995, Lovins had realized that Hypercars would kill the oil industry: "The Middle East would therefore become irrelevant and the price would crash. With so little demand, most of the oil in the ground would be no longer worth extracting."

And by mid-1998, as Lovins contemplated the switch to hydrogen-fuel-cell technology, he realized it could completely displace the coal and nuclear industries as well, as he wrote in a letter to *Science* magazine:

> Ultralight hybrid-electric cars have multi-billion-dollar private commitments, are coming quickly to market, and will ultimately save as much oil as the Organization of Petroleum Exporting Countries now sells. The most efficient will use H_2 fuel cells whose immediate commercialization, now feasible, can displace most if not all oil, coal, and nuclear power at a profit.

By May 2000, Lovins had expanded his list of industries that the Hypercar would impact. It would bring about the "end as we know them" of the automobile, oil, steel, aluminum, coal, nuclear, and electricity industries.

And it would not take long to bring these industries to their knees because, as Lovins put it, "Hypercars will be widely available in about five years [2006], dominant in about ten years [2011], and the old car industry will be toast in twenty years." At first I was puzzled by the disappearance of the electricity industry, but, of course, Hypercars were to replace most of the large power stations by generating electricity from hydrogen when parked at work and at home.

The Hypercar Fallacy: "1 plus 2 equals 10"

Lovins believed that combining two efficiency ideas would get us more than the sum of their savings, and called this his "1-plus-2-equals-10-equation." Let's check the math.

First efficiency idea: using carbon fiber: GM's carbon-fiber Ultralite got 88 miles per gallon. That's about four times better than normal cars were getting. For round numbers, say the "carbon" idea takes a normal car from 20 to 80 miles per gallon.

Second efficiency idea: hybrid motor: This is the idea Lovins combined with carbon to come up with the Hypercar. Using both ideas, he claimed that a "300–400-mpg four-seater with widely available technology was possible." For round numbers, say the hybrid idea takes a carbon car from 80 to 320 miles per gallon. That's four times better than a carbon car and sixteen times better than a normal car. These claims are exaggerated, but bear with me.

So a four-times-better idea (carbon) combined with another four-times-better idea (hybrid) is sixteen times better. So in this example, "4 + 4 = 16." That's what Lovins meant by "1 plus 2 equals 10." The miles per gallon more than add up.

But the idea is to save gasoline, so we had better check gas savings.

Suppose the normal 20-mpg car used 800 gallons in a year. Then the carbon car would use only a quarter of this, or 200 gallons, and it would save 600 gallons a year.

Similarly, adding a hybrid motor to a normal car would quadruple the miles per gallon and save 600 gallons a year.

But if a hybrid motor is added to a carbon-fiber car, what happens?

Intuition: Because a carbon car only uses 200 gallons, there is no way adding a hybrid motor can save 600 gallons.

Math: Adding a hybrid motor to a carbon car cuts gas usage four times, from 200 to 50 gallons, for an additional savings of 150 gallons and a total savings of 600 + 150, or 750 gallons saved.

So a 600-gallon idea (carbon) combined with another 600-gallon idea (hybrid) makes a 750-gallons-saved idea. In this example, "600 + 600 = 750."

The truth is that the hybrid-motor idea saves much less, not much more, when applied to a super good car like GM's Ultralite instead of to an ordinary car. This is well known, and it's why GM never "thought of" adding a hybrid motor to a carbon-fiber car.

In November 2000, as Lovins explained, Hypercar Inc. had "developed for a few million dollars in 8 months, on time and on budget," the first show-car version of the Hypercar, which they dubbed the Revolution.

The Last of the Hypercars

While realistic in appearance, the Revolution show car lacked a carbon-fiber body, lacked a motor, and had no hydrogen fuel cell and no fuel tank. The car was not full sized—it was just for show. Amory Lovins never did get to drive a Hypercar. In 2004, Hypercar Inc. changed its name to Fiberforge and stopped trying to convert the world to Hypercars.

In 2007, California's Air Resources Board estimated that hydrogen fuel-cell vehicles might become widely available in 2025—twenty years later than Lovins predicted. And they would still not include the central Hypercar design feature, a carbon-fiber body. Plug-in hybrids, however, are currently looking more promising than hydrogen cars, so Hypercars may never get their day in the sun.*

What's Wrong with a Little Optimism?

Optimism can inspire action, but it should not cloud our vision. Believing that Hypercars will end oil addiction and ward off climate change can make policy-based approaches seem unnecessary. As Lovins says, "Growing evidence suggests that besides fuel taxes and efficiency regulations, there's an even better way: light vehicles can become very efficient through breakthrough engineering."

In other words, Lovins is saying his "better way" makes energy-efficiency policies unnecessary. But Lovins proved, by rigorous experiment, that this "even better way" is next to impossible. He had a better chance than anyone of finding it, and he did his best for nearly fifteen years. In the end he could not get a single prototype Hypercar produced.

It may have been what Lovins calls "cultural barriers"—in other words, a lack of faith by others in his concept—but if so, Lovins saw this from the start and did his best to breach those barriers. On the other hand, it may have been that the Hypercar was just too expensive, as industry leaders apparently decided. If so, Lovins has demonstrated that betting on breakthrough technologies is far too risky, even when the world's leading energy guru places the bet.

Perhaps Lovins claimed that government policies were unnecessary only as a way of promoting the virtues of his Hypercar. Perhaps he didn't mean it. But Lovins's paper "Four Revolutions in Electric Efficiency" seems to confirm his dismissal of efficiency regulations. It's like the Sherlock Holmes mystery about the dog that didn't bark when a crime was committed. In twenty pages, his paper contains no hint of appliance standards, even though that's the first thing one would have expected.

In 1978, California passed the first refrigerator-efficiency standards. The federal government followed in 1987, scheduling eight appliance standards, including refrigerator standards, to take effect on January 1, 1990. This was the most publicized and most high-impact electric-efficiency event ever, and it was in progress while Lovins wrote his article on electric efficiency. Why would Lovins fail to mention it?

In effect, his article argues that we do not need standards because the four revolutions he sees happening on their own will be vastly more effective. The nicest thing I can find Lovins saying about building codes and appliance standards is that they are "better than nothing."

～

The trouble with Lovins's optimism is that it is not just a little optimism. It overwhelms all other approaches. It says we don't need efficiency standards or really any government policies. All we need to do is wait to buy a Hypercar and keep an eye out for new efficient technologies that will save us money. New technology will crush OPEC, the coal industry, and the nuclear industry. Global warming will fade away.

Lovins is right to favor conservation and right to favor the use of markets. Some of his ideas are practical. But three centuries of technical progress have brought unimaginable efficiency gains—and vastly increased use of fossil fuel— without solving our energy problems. Something more is needed than Lovins's promise of "breakthrough engineering" and faith that corporations will break down their "cultural barriers." Lovins's objectives are well intentioned, but his hyper-optimism is a barrier to almost every effective energy policy.

chapter 6

No Free Lunch?

Increased fuel efficiency, however, is not free. … Any truly cost-effective increase in fuel efficiency would already have been made.

—Former Council of Economic Advisers Chairman
N. Gregory Mankiw, 2007

N. GREGORY MANKIW IS THE MIRROR IMAGE of Amory Lovins, the protagonist of Chapter 5. Lovins knows that every energy measure we could possibly need will save more than it costs. Mankiw knows that all such measures will cost more than they save. Mankiw served as George W. Bush's chairman of the Council of Economic Advisers from 2003 until 2005 and is well respected within the economics profession.

The Mankiw-Lovins bipolarity highlights an important split in energy policy circles. On one side, we find Mankiw and other "neoclassical" economists. They oppose not only fuel-economy standards but all energy-efficiency standards and energy-efficient building codes. That is, they oppose all measures favored by the "physics camp" that I mentioned in the last chapter.

On the other side of the split, the physics camp is less strident. Although they tend to believe efficiency standards are most important, they rarely take a strong stand against the policies favored by the neoclassicals. Although Lovins shares the camp's belief in abundant, cheap efficiency measures, he is not typical of the physics camp, because he sees less need for standards than do most in that camp.

The neoclassicals, being economists, favor policies that change the price of energy. They call this "sending a price signal" to the market. They favor sending the price signal by taxing fossil fuel. That would, of course, raise its price. But since taxes are unpopular they've come up with a stealth tax, which is not so easily recognized. It's called a "cap-and-trade" policy, and six or seven of these have now been proposed to Congress. Like a tax, a cap-and-trade policy raises the price of fuel and electricity. It "sends a price signal," which pleases the neoclassicals.

The physics camp tends not to like either taxes or stealth taxes, both of which they see as unpopular because they are clearly costly, not cheaper than free. Instead, they prefer to mandate more efficiency with a standard, which they have precalculated will save more than it costs. While working at Lawrence Berkeley National Laboratory, I helped make these precalculations for national appliance standards.

So where do "free lunches" fit into this controversy? "A free lunch" is what the neoclassicals call any policy that provides a benefit that is greater than its cost. The term is descriptive, but it also conjures up the slogan "there's no such thing as a free lunch," which helps them win their point.

In summary, the neoclassicals say: The physics camp claims all its proposals are free lunches, but there's no such thing—we need taxes. The physics camp says: Call them free lunches if you like, but there are a lot of ways to save money and energy at the same time—who needs taxes?

The Energy Policy War

The neoclassicals dismiss the efficiency programs of the physics camp saying they are not free lunches; they cost more than they save, and that's a waste of money. But there is also a net cost to the efficiency gains from the neoclassicals' taxes, so why is one better than the other?

The neoclassicals reply that neither approach provides a free lunch, but that their approach provides cheaper lunches than the physicists' approach. They propose sending a price signal to the market and letting the market choose how to improve efficiency. Since, by assumption, markets always do better, the physicist proposals are always worse. So say the neoclassicals.

Because the neoclassicals see taxes as a more market-based approach and markets as better than government, they actively oppose all efficiency standards. Under the administration of George W. Bush, the neoclassicals helped to derail appliance-efficiency and fuel-economy standards—government-run, free-lunch programs all.

William Nordhaus, a Yale economist who has probably spent more time studying energy and climate-change policy than any other economist, simply calls such policies "fluff." Lovins is well aware of this view and enjoys talking about putting "several trillion dollars back in Americans' pockets" and then saying, "That's not a free lunch. It's a lunch you're paid to eat."

Mainly, the physics camp spends its time defending its position rather than attacking the neoclassical position. But since the times when a Democratic Congress attacked President Gerald Ford's $2-per-barrel tax on imported oil, the physics camp, by constantly downplaying the importance of energy prices, has lent support to those who oppose strengthening market-oriented price signals. Many in the physics camp believe in so many free lunches that they think price matters very little.

I am particularly interested in this policy war, because I believe one key to recovering from oil addiction and reducing carbon dioxide emissions is a fuel-economy policy for cars and light trucks (including SUVs), something the neoclassicals dismiss out of hand. At the same time, I believe the neoclassicals are right that raising the price of carbon is the most important step, though I favor doing so with an untax, not a tax. The policies of the two sides, in my opinion, actually complement each other.

Efficiency Measures Can Save Money

Economists are not all strict neoclassicals. Allow me to introduce Stanford economist Kenneth J. Arrow. A winner of the Nobel Memorial Prize in Economic Sciences, Arrow is one of the most respected of all economists and is a central figure in the development of mathematical neoclassical economics. In 2007, after the UK government issued a major report estimating that the cost of climate stabilization would be between +3.4 percent and −3.9 percent of the world's total gross national product (GNP), here's how Arrow responded: "Since energy-saving reduces energy costs, this last estimate [negative 3.9 percent] is not as startling as it sounds."

If Arrow thought a cost of negative 3.9 percent of GNP was impossible, he would have called it startling. Instead, he said it was not so startling. Arrow is saying that a cost estimate of negative 3.9 percent of GNP just might be right. A negative cost means a net savings. Since this concerns the global economy, he is saying the world just might find $3 trillion per year of free lunches from energy-savings schemes, if it went looking.

In other words, according to Arrow, quite a few of the physicists' favorite policies *might* save more than they cost. This opens a door slammed shut by the neoclassicals' extreme views, which

Suspect from the Start

The *Journal of Economic Literature,* perhaps the most prestigious of economic journals, published an article on the rational consumer part of neoclassical theory in 2002. According to the authors, economist Paul Samuelson proposed the neoclassical theory of extreme rationality concerning future costs and savings in 1937, and people quickly accepted his idea because of its convenience and simplicity and in spite of Samuelson's reservations about its accuracy. In other words, economists began using the theory without testing it.

The article goes on to explain that once economists began checking the theory, the "empirical research led to the proposal of numerous alternative theoretical models," none of which agree with the neoclassical theory of extreme rationality. The article reports on dozens of papers, with dozens of empirical results, most of which contradict the neoclassical theory. People, it seems, may not always be entirely rational, even about money.

are based on an assumption of completely rational consumers. However, it's important to note that both Arrow and the author of the report believe that total cost is more likely to be plus than minus. That means they believe that, although some policies may save more than they cost, on average the policies will cost more than they save.

If it just might be possible to save a few trillion dollars per year instead of paying a few trillion extra, it seems foolish not to even try just because of some disputed economic theory (see "Suspect from the Start").

The Taste of a Free Lunch

When Art Rosenfeld looked into refrigerator efficiency, he didn't need any fancy economic theory to tell him we were being charged way too much for "lunch." Art Rosenfeld is a real physicist. He coauthored a text in nuclear physics with Enrico Fermi, who developed the first nuclear reactor and who won the Nobel Prize in physics in 1938. Rosenfeld also participated in the discovery of subatomic particles with Luis Alvarez, who won the Nobel Prize in 1968. In 1973, at the start of the first energy crisis, Rosenfeld noted that "if we Americans used energy as efficiently as do the Europeans or Japanese, we would have been exporting oil in 1973." He's been the country's top energy-efficiency expert ever since.

By 1975, Rosenfeld was hard at work developing residential building standards, and in 1976 he recommended an efficiency standard for refrigerators and freezers to California's governor, Jerry Brown. That's how appliance standards got started.

When Rosenfeld looked into refrigerator efficiency, he found a wide range of efficiencies but no correlation at all between cost and efficiency. It appeared that a lot of money could be saved on electricity by buying an efficient refrigerator that didn't cost any more—but people weren't doing that.

Before standards, manufacturers skimped and used fiberglass insulation instead of rigid polyurethane. They made the walls thin to get more room inside. With thin walls and poor insulation, the outsides of the refrigerators got cold enough in spots to cause condensation. To prevent this, some manufacturers installed heaters in the outer walls of refrigerators! The heater uses energy, and then the refrigerator uses more energy to cool the heater.

Apparently, because consumers paid no attention to efficiency, manufacturers saw no point in spending much money to make them efficient. This was reflected in the history of refrigerators. Between 1950 and 1974, energy use per refrigerator grew more than twice as fast as refrigerator size. While the size of refrigerators more than doubled, their energy use more than quadrupled.

But the high energy prices of the first energy crisis changed all that. People starting thinking about saving energy, and that's difficult when you have no idea how much energy an appliance uses. So in 1975, the federal government required energy-efficiency labels on some appliances. In 1978, California imposed efficiency standards, tightening them in 1981 and 1987. The federal government took over the process and set even tighter standards in 1990 and again later.

As a result, by 2001, refrigerators used 69 percent less energy than in 1974 even though they were 20 percent larger. Saving that much electricity saves $127 per year. Meanwhile, the cost of a refrigerator had dropped by half. If the extra efficiency had a cost, it could not have been much because by 2001 the average price of a refrigerator was only $850. Even if $400 of that was attributable to efficiency, which is highly unlikely, it would have been repaid in three and a half years by the reduced cost of electricity. For the next twelve or so years a refrigerator lives, the $127 per year of energy savings would be gravy. It looks like refrigerator standards are a lunch we're paid to eat.

Are New Car Buyers 100 Percent Rational?

I am skeptical that Mankiw read any studies (if they exist) proving that car buyers are rational on average before he predicted in the *New York Times* that "any truly cost-effective increase in fuel efficiency would already have been made." Neoclassicals usually rely on their theory for these sorts of pronouncements.

Cars now come with Environmental Protection Agency (EPA) mileage ratings, but these fall far short of telling consumers their total future gas costs if they buy the car. Neoclassical economics assumes consumers are quite good at estimating future costs. In particular, neoclassical theory assumes consumers will

- ▶ Estimate the price of gasoline for the next ten to fifteen years.
- ▶ Estimate how many years they will keep their cars.
- ▶ Receive the full remaining value of gas savings when they sell their cars.
- ▶ Estimate how far they will drive their cars each year before selling them.
- ▶ Estimate how much their actual mileage will deviate from the EPA ratings.
- ▶ Discount future savings at a percentage corresponding to either the interest rate on their credit cards or the interest they earn on investments. (Even economists find this one confusing.)

If consumers make all of these estimates without bias and purchase their new car on this basis, Mankiw should be right. Most fuel-efficiency experts see it differently. They believe consumers take account of less than half of a car's future gasoline costs.*

Limits to Free Lunches

So it looks like the neoclassicals are wrong. Call them free lunches if you like, but there are opportunities to save more on energy than it costs to gain efficiency. And, at least in the case of refrigerators, it looks like some of that opportunity was captured by a government regulation—an efficiency standard for refrigerators.

But all this really shows is that the extreme neoclassical position is wrong. Perhaps very few efficiency standards can save more than they cost, or perhaps the opportunities are enormous. Either the neoclassical view or the physics camp could be nearly correct.

Unfortunately, both sides are so sure they are right that neither side documents their case carefully. Even the case for refrigerator standards is clear only because it is dramatic, not because it is well documented.

Generally, claims that efficiency programs will save more than they cost omit four considerations, each of which can be quite important:

- ► Regulatory inefficiencies.
- ► The take-back effect.
- ► Consumer inconvenience.
- ► Consumer variability.

Regulatory Inefficiencies. Neoclassicals assume perfectly efficient markets. Physicists implicitly assume perfectly efficient regulation. This bias is the result of omitting any cost for regulatory mistakes, such as setting a standard incorrectly. I have not found energy regulators to be any more rational than new car buyers.

The Take-Back Effect. When an appliance is made more efficient, it often becomes cheaper and more convenient to use. Consequently, people use it more or buy a bigger one. This is a benefit to society and actually makes efficiency programs more valuable than the physicists claim. But it also means efficient appliances use more energy than physicists estimate.

Consumer Inconvenience. Some ways of gaining efficiency cost no money, but do cause inconvenience. For example, making the walls of a refrigerator thicker means it either takes more space in your kitchen or holds less food. The cost of such inconvenience is nearly always ignored.

Consumer Variability. If I run my air conditioner 1,000 hours per year, any improvement in efficiency will be 100 times more valuable than if I run it 10 hours per year. It does not make sense for a low-use appliance owner to buy as much efficiency as a high-use owner. This means that even the best efficiency standard is likely to be a waste of money for the low-use owner. I have never seen this accounted for.

Although there are many imperfections in markets, that does not mean there are many free lunches. There's only a free lunch if the problem can be fixed at a cost that is less than the savings. All four of the considerations just discussed either raise the cost of fixing the problem or reduce the benefit. Because they are generally ignored, the claims of free lunches are frequently overstated.

A Pricing-versus-Efficiency Compromise

The physics camp wants many efficiency regulations and cares little for price signals. Neoclassical economists want only price signals and no efficiency regulations. The resolution of this conflict flows from the first principle of fossil philosophy, as explained in Chapter 1: *Treat the problem, not the symptom.*

The two camps focus on two different problems. The price of fossil fuel is too low, so we need the neoclassical solution of higher price signals. Consumers are shortsighted when evaluating future energy savings, so some efficiency standards can help them save money. Many if not most economists favor both approaches when each is used to solve the matching problem.

This compromise rejects the extreme neoclassical position, but it also requires two changes in the tendencies of the physics camp. It requires taking fossil-fuel prices far more seriously, and it requires backing away from the notion that physicists know how to fix literally hundreds of market imperfections while saving money.

I believe economists are right to be suspicious of large numbers of "market-fixing" efficiency regulations. Their skepticism is not based on an implausible assumption of consumer rationality, but on the four realistic concerns listed above, which are consistently ignored by the physics camp. Moreover, the design of even major efficiency standards is poor and fails to use modern economic tools. This has resulted in such fiascoes as fuel-economy standards that remained at their initial 1975 setting for over thirty years and that reward designs that kill more people while using more fuel. I am referring to the requirement that cars be aggressively redesigned so they can be reclassified as trucks and qualify for a lower fuel efficiency standard.

It would accomplish far more to design the major standards well and evaluate them carefully rather than to charge ahead with hundreds of smaller measures that ignore economic concerns. But the real challenge for the physics camp is to accept the importance of price and to realize that their entire campaign is at risk without the proper price signals.

Take-Back by the Numbers

Here's how the evaluation of savings from compact fluorescent bulbs can go wrong. The root problem is the take-back effect. But evaluators amplify the problem by basing calculations on new bulbs and not on the ones replaced (because they don't know what was replaced).

- Replace a 40-watt incandescent bulb with a 100-watt-equivalent compact fluorescent lamp (CFL).
- The CFL uses 23 watts.
- The actual savings is 17 watts.
- CFL program evaluation assumes that when a 100-watt CFL is used, it replaces a 100-watt incandescent bulb (they can't tell, so they make this guess).
- Replacing a 100-watt bulb with a 23-watt bulb saves 77 watts.

Calculated savings: 77 watts. Actual savings: 17 watts.

In addition, because the light is cheaper, people may leave it on more, and the savings could actually be zero. This is an extreme case, but it happened in my kitchen. Most efficiency gains are not lost to the take-back effect—at least not right away.

Having worked in the physics camp for years, I have heard many excuses for ignoring the take-back problem but have never actually seen it taken into account. This is too bad because, if the problem were faced squarely, the natural conclusion would be that the neoclassicals' price signals do not substitute for efficiency measures but rather are a necessary complement.

Think about the history of take-back. Ordinary lightbulbs are 150 times more efficient than candles. But we don't use 150 times less energy for light, we use more energy than in colonial times. Scientists of the past have provided us with enormous efficiency gains but never enough to reverse our increasing use of energy and fossil fuel.

So the physicists are taking a real gamble. Compact fluorescent bulbs may save energy this year, but ten years from now people may have discovered they can afford to light their gardens at night as brightly as the sun lights them in the daytime. And the less energy a light uses, the less it pays to switch it off. If history is a guide, increasing wealth combined with the take-back effect will eventually win out over the energy savings of increased efficiency.

There is a simple way out of this dilemma. Raise the price of electricity and refund the extra cost. This is again the untax, and exactly why this works will be explained in Chapter 16, but here is the outcome. The higher price of electricity reduces or reverses the take-back effect. High energy prices, even with a full refund of the energy tax revenues, will greatly reduce the risks of take-back. Then increasing efficiency will work as promised.

～

The compromise between the extreme neoclassical camp and the physics camp is simple and constructive. The most important efficiency programs, especially fuel-economy standards, should be accepted and perfected. Neoclassical economists should stop arguing against these on the basis of untested theory.

The physics camp should recognize that there are real problems with "fixing" markets and that blind faith in regulatory fixes is no more appropriate than blind faith in markets. Physicists and economists should join forces to make the big efficiency programs work better and to implement better fossil-fuel prices. This will help protect energy-efficiency gains from the take-back effect.

chapter 7

The Core Energy Plan

The entire carbon tax should be returned to the public. … Carbon emissions will plummet far faster than in top-down or Manhattan projects.

—James E. Hansen, NASA climate scientist, 2008

PREVIOUS CHAPTERS DISCREDITED THESE MYTHS: that we will wreck the economy, that peak oil will herald doom, and that miracles are imminent. Other chapters explored why it is foolish to ignore climate change or shun money-saving policies. Leaving these misconceptions behind, I will now sketch a Core National Energy Plan that is cautious yet powerful.

Part 3 of this book lays out details of the plan. So if you find the workings of the untax, or the race to fuel economy, a bit puzzling, don't be surprised. There are a few tricks to good economics, and the full explanation will make more sense after a closer look, in Part 2, at how energy markets work.

The core energy plan flows from basic principles. A good design does not rely on incredible advances in technology. Instead, a good design requires that a plan be

- ▶ Simple.
- ▶ Cost effective—a bargain.
- ▶ A treatment for the disease, not just for the symptoms.

Simplicity helps prevent mistakes and gaming. I have learned this repeatedly in my work diagnosing and adapting electricity markets. I have also learned that this principle is seldom respected in practice. But simplicity is still the right way to begin.

Asking for a bargain may seem superficial, but, in fact, that is exactly what economists mean when they call for "efficiency," their primary objective. The cost of saving a certain amount of oil or carbon should be as low as possible.

Unhealthy energy markets—ones that are inefficient and do not reflect social costs—develop symptoms such as gas-guzzling cars, too few wind turbines, and too many coal plants. The symptoms are the ways energy is wasted. The underlying disease involves "market failures"—basic problems with how the market works. Treating the symptoms—for example, by subsidizing ethanol—often causes unwanted side effects. And there are just too many symptoms to treat them all one by one. A better approach is to identify underlying causes—aspects of the market that are broken—and treat those rather than the symptoms.

Energy Policy: Mostly Sound and Fury

Yale economist William Nordhaus, writing in the *New York Times* in 1980, had this to say about fixing the cause of the problem:

"A recent study by the Department of Energy, called Energy Programs/Energy Markets, has estimated … what the impact of all current programs would be in 1990. … The central and surprising conclusion of the Energy Department study is that the energy programs add up to about zero. … By comparison, the rising relative prices of energy will probably lower energy use 20 to 30 percent by 1990."

What's Broken?

To avoid treating symptoms, we must identify the problems. Almost everyone has a list of things they find wrong with the market, so the trick is to decide which are worth fixing. Amory Lovins, the lead optimist in the physics camp, sees market barriers by the dozen and urges us to "clear them," "bust them," and "vault over them." Market "barriers," or "failures," as economists call them, are broken aspects of markets, such as landlords who buy inefficient appliances for tenants because the landlords do not pay the electric bills. I believe most economists are open to the idea that many little things go wrong with markets, but they take a cautious view of such problems.

Having seen many proposed and attempted market "fixes," economists tend to shy away from jumping on the fix-it bandwagon. Market fixes usually come with their own problems, and for minor market failures the cure is usually worse than the disease. Economists recommend identifying the worst problems and focusing policies only on those few. A good solution to an important problem puts us well ahead of a multitude of poor solutions to lesser problems. William Nordhaus identified the shortcomings of piecemeal policies in 1980 (see "Energy Policy: Mostly Sound and Fury").

The four most important energy market failures are listed below. Although the idea of consumer myopia, discussed in Chapter 6, has less backing than the others, I believe most economists will agree that the following are the energy market problems that deserve the most attention.

The Four Energy Market Failures

▶ **Omitted negative side effects of fossil fuel.** (1) Environmental costs of pollution and CO_2 emissions. (2) The costs to the United States and its allies to secure uninterrupted oil supplies.

▶ **Market power.** The Organization of Petroleum Exporting Countries (OPEC) overcharges for oil relative to the competitive price.

▶ **Consumer myopia.** Consumers see future energy costs unclearly and react to them less than to a product's purchase price.

▶ **Omitted positive side effects of advanced research.** Discoverers of basic new technologies are under-rewarded.

The first problem with the market is its failure to take into account the costs that are external to the market. These are the costs of the environmental side effects and the costs of securing oil supplies. There are many costs in each category, but I will lump those in each group together. For environmental costs, I will focus only on the cost of CO_2 emissions—the main driver of global warming. (Economists call *side effects* "externalities.")

The cost of securing oil supplies points toward a policy of using less oil and so does the second market failure—OPEC's market power. Some will question the extent of OPEC's power in recent years, and this will be discussed in Chapters 13 and 29. But the policy I will propose will serve to lower the world price of oil in any case, so there is no harm in assuming OPEC still has a lot of power.

The third market failure, consumer myopia, is the tendency of consumers to underestimate future energy costs and buy energy-inefficient products. The fourth problem with the market is that fundamental research is risky, and the benefits from a breakthrough may be much greater than the reward. This leaves fundamental research insufficiently rewarded by the market.

What's the Plan?

A simple four-policy national energy plan is all we really need. Of course, there is room for add-ons, but four basic policies are essential and would do far more than we accomplish now. I will focus on the first three policies of the plan, as these are the least understood. The fourth policy is simply to fund more basic research.

The Core National Energy Plan:

- ▶ An untax on carbon.
- ▶ A separate untax rate for oil carbon.
- ▶ A carmakers' race to fuel economy.
- ▶ Public funding of basic energy research.

As good market design requires, the plan is simple. Because it respects competitive market principles, it's also a bargain. As we'll see in a moment, it saves money by harnessing the ingenuity of every American—from CEOs to high school students. It's also fair in that it rewards all those who help out, and to the extent the poor use the least energy, it rewards the poor for doing so.

The next three sections explain the first three policies of the plan, beginning with the "untax," which raises no revenues for the government, but refunds all revenues to consumers. After introducing the untax, I explain why the untax rate for oil carbon should take account of OPEC's oil prices. Finally, I explain the race to fuel economy, which is more fair to car companies than standards and can be as powerful as desired.

Meet the Untax

"Among policy wonks like me, there is a broad consensus. ... If we want to reduce global emissions of carbon, we need a global carbon tax." So said Mankiw, whom I disagreed with over fuel economy in the previous chapter. I agree completely with Mankiw on this—the central point of his article in the *New York Times*.

Mankiw says there is no disagreement "between environmentalists and industrialists, or between Democrats and Republicans" on the benefits of a carbon tax. He's right. A carbon tax is the cheapest way to solve the first, and most important, energy market problem, "the missing cost of carbon emissions."

But as Mankiw also reminds us, both American voters and political consultants consider "tax" a four-letter word. Can we find a way around the political lightning rod of "taxes" to save Americans tens of billions of dollars a year by implementing the best energy policy?

Mankiw comes close to finding the way. There are two halves to any tax—how it is collected and how it is spent. The benefits of the carbon tax come entirely from the first half—the charges on carbon, which increase its price and makes us all look for ways to avoid using fossil fuel. So economists look for ways the government can spend the tax revenues to make voters happy. Happy enough to forget it's a tax? Not likely.

Mankiw proposes to spend the carbon-tax revenues on a "rebate of the federal payroll tax on the first $3,660 of earnings for each worker." That is close to the right answer. Others propose reducing income taxes, either personal or corporate, and some propose spending it on research and subsidies.

To find the right answer, we must go north to Alaska, where it was discovered in 1976. The answer—how the government should spend the money—couldn't be simpler. Don't spend it! Just give it back to us, thank you very much. Alaska sends identical checks, for about $1,000, to every Alaskan resident every June. It collects these revenues from its famous oil pipeline. This is popular. This is the key to an untax.

Taxes raise money for the government. The office football pool collects money and gives it back. That's not a tax. That's an incentive to correctly predict the winning team. It's also fun.

A carbon untax is an incentive to use less carbon. Use the average amount of carbon, and your refund check will exactly cover what you contribute indirectly to the carbon pool of money collected from oil, gas, and coal companies. These companies will tell you how much you're contributing, but they will raise prices to cover their carbon charges from the untax. That's exactly what's needed to discourage the use of fossil fuel.

Use more carbon than average, say by flying your own personal jet, and you will pay more in higher fossil prices than you get back in June. Because the rich tend to use far more than average, 60 percent of us are actually below average and will get back more in June than we pay the rest of the year in higher fossil prices. The less carbon you use, the greater your winnings. Or, if you fly your own jet, the less carbon you use, the less money you lose. That's why, even though it gives back all the money, the untax works perfectly. Chapter 16 gives the full explanation.

Charging OPEC

The second policy in the plan specifies a separate untax rate for oil. When OPEC pushes the price of oil high enough, that in itself is a strong global warming policy (see Chapters 2 and 8). There is no need to raise the cost of oil still further, so when the oil price is high enough, the carbon charge on oil should drop to zero.

For example, when the world price is $80, the untax might be $20, but if the world price rises to $100, the untax rate would fall to zero. The sum of the world oil price plus carbon charge paid by refineries would be $100 either way. This price stability protects alternative fuel investments, such as those in advanced ethanol plants and investments in conservation such as hybrid or electric cars. Investors worry that the price of oil may collapse and leave their investments worthless. This happened in 1986. OPEC has even threatened to do this deliberately in order to discourage energy investments that would reduce our addiction.

With a variable oil-carbon charge and an untax, if OPEC lowered the world price of oil for a couple of years, the carbon charge would rise to keep the domestic price of oil high. This would protect alternative fuel suppliers, and consumers would still capture the benefits of low world prices through higher untax refund checks.

As explained in Part 4, an untax on oil is the right basis for a consumers' cartel, and as such it's an incentive for international cooperation. This is particularly true for China, which will soon be even more addicted to oil than is the United States. A successful global warming policy requires such international cooperation, especially from China, so the untax on oil serves both goals—climate stability and energy security.

The Race to Fuel Economy

In 1975, Congress set the Corporate Average Fuel Economy (CAFE) standard for 1985 cars at 27.5 mpg. In 2010, the standard will still be 27.5 mpg. Once high OPEC prices started coming down in the early 1980s, the CAFE machine just stopped percolating. After the return of high oil prices in 2006 and 2007, Congress passed legislation that requires the standard to increase to 35 mpg in 2020. However, the bill requires nothing until 2011, and then only at the discretion of the president. The risk remains that if oil prices drop, the standards may end up lower or go into effect later, as happened in the mid-1980s.

CAFE standards have two fundamental flaws: The Big Three automobile manufacturers hate them, and they are easy to gum up. The two flaws work together all too well. As soon as the country settles down after an OPEC crisis, the Big Three gum up the standards. No good reason exists for such poor design, and after thirty-two years it's time for a change.

No one would think of requiring athletes to perform to standards at the Olympics. No one wants government standards saying how tasty the food should be at their favorite restaurant. Athletes compete. Restaurants compete. Car companies compete on everything else but fuel economy—the one thing they do poorly at. Competition is not a new idea, except to regulators.

Chapter 20 explains how to turn CAFE standards into a competitive race to fuel economy in which losers pay for the prizes. The race mechanism eliminates standards entirely; each company simply tries to do better than the others. The better it does, the greater its prize (or the less it contributes to prizes for others). With a standard, companies lose the incentive to keep trying once they reach that standard.

To keep the Big Three happy, I will suggest rigging the race in their favor a bit. Even so, every car manufacturer will get the same reward for each extra bit of fuel efficiency, so they will all try equally hard. The incentive can be set just as strong as we want by adjusting the prize.

Also, there is no need to delay the start of a race for four years, as our government just did again with CAFE standards. All car companies can do their best, whatever that is, the very first year. Incidentally, similar legislation can make appliance standards more effective and vastly simpler.

How Much Does It Cost?

The first three policies of the Core National Energy Plan are all revenue-neutral. The two untaxes pay back to consumers exactly, to the penny, what they collect. The Department of Energy pays an administrative cost, but in Alaska this amounts to less than 1 percent of the income distributed. The third policy, the fuel-efficiency race, simply redistributes funds from losing car companies to winning car companies. The last policy, public funding of basic energy research, is fairly cheap. We can beef up the research budget for conservation and non-nuclear alternative energy by ten times, and it still comes to only about $10 billion a year, which is one-fifteenth of 1 percent of the gross domestic product.

Does being revenue-neutral mean the first three policies are free? No. Although an untax refunds all the money it collects, it still involves the indirect *net* costs that consumers incur to reduce their energy use. Net costs are small because they are the difference between the cost of saving energy—for example, buying a hybrid car—and the value of the energy saved. Because saving energy is voluntary, people do not choose to spend much more than they save. The economics of net costs will be explained in Part 3, but one point is most important and simple to grasp.

Revenue-neutral policies come with a sort of guarantee: If they don't work, at least they entail no net cost. That's because a revenue-neutral policy refunds all taxes or fees, and if we don't respond and do something to save carbon, we incur no indirect cost.

Also, because we are careful about how we save energy, indirect costs are relatively low. As an example, suppose the untax collects $300 billion and refunds it all and that saves 20 percent of our carbon (a good start). Using the economics explained in Chapter 16, the net cost to consumers will be only about $38 billion.*

If the race to fuel economy is designed correctly, it will have a negative net cost. The efficiency race is only intended to solve the third energy market problem, consumer shortsightedness. If it does that, and no more, it will save consumers more money on fuel than it costs them for efficient cars. I will not attempt to estimate the net savings, but consumers spent roughly $300 billion on gasoline in 2006, leaving room to save real money.

Excluding net savings from the fuel-economy race, the total cost of the Core National Energy Plan comes to about $48 billion per year. This is only one-third of 1 percent of the national income—wait four months and we will be that much richer from normal economic growth. This would be a vastly stronger policy for both energy security and climate change than what we have now. When world oil prices are high, the cost of the policy would be considerably less because the carbon charge on oil would be low and perhaps zero.

Can We Charge It to OPEC?

Based on a 20 percent cut in U.S. oil use, the world price of oil would be reduced by about 6 percent, making OPEC and Big Oil together pay roughly $26 billion of the cost of these policies. This assumes that oil will cost $75 a barrel without an energy policy. But the full proposal of this book calls for an international consumers' cartel to challenge OPEC, which is the international producers' cartel. Such a consumers' cartel would at least double the savings for the United States. This would cover the full cost of this sample core energy plan.

By 2050, if climate change policies are ramped up to the level that is frequently anticipated as necessary, their cost would likely outstrip the savings from reductions in the world price of oil. But that, of course, depends on how much cheap conservation is available, future technological breakthroughs, and how short of oil we would be without a strong climate policy. But for the next ten or twenty years, we can charge it to OPEC and Big Oil.[1]

Will the Core National Energy Plan Work?

The untax is at the heart of the policies I propose. Will the untax work? First, as Mankiw points out, the idea is close to a century old and trusted by more economists than any other approach. Second, this is very close to the policy tested by OPEC, and it passed with flying colors. It stimulated a huge amount of conservation and a significant increase in supply. It reduced carbon dioxide emissions from the United States, and it crushed OPEC's price for eighteen years. OPEC put a charge on oil, just the same as the untax, but forgot to put our refund checks in the mail.

A $300 billion untax would mean a $1,000-per-person refund every year. Because, it's a more balanced approach, targeting all fossil fuels and not just oil, an untax would accomplish more at less cost than OPEC's approach.

A family of four that changed from using 50 percent more carbon than average to using 25 percent less than average would save $3,000. This is a strong-enough incentive to cause people to buy better lightbulbs, more insulation, and less thirsty cars. Businesses will have the same-strength incentive because they save the same amount when they use less fossil fuel.

---

1. Chapters 13 and 29 provide a complete discussion, and a box at the end of Part 4 provides a example calculation showing that we could charge it to OPEC.

The strength of the untax is the breadth of its reach. Subsidies require regulators to target particular carbon-saving methods, and even emission caps are only half as broad as an untax. The untax targets every carbon-saving method that 300 million Americans can dream up. This is the strength and beauty of a true market approach. It harnesses the creativity of every entrepreneur, inventor, high school student, and parent. It motivates the rich and the poor alike. It stimulates car pools, neighborhood organizations, citywide efforts, and state programs. It promotes innovation at national laboratories, huge corporations, and little alternative energy start-ups. And because the untax treats all equally, the best ideas win out.

Compared with such a massive and balanced approach, specialized approaches that target things like corn ethanol, hydrogen cars, wind turbines, or solar roofs hold little promise. In fact, the untax would appropriately reward the users and developers of each of these technologies and allow the market to select the real winner among them—if there is one. Compared with choosing technologies in the dark, according to which is backed by the strongest congressional lobby, the untax is like the light of day.

Don't Touch the Untax

I end this chapter with a strong warning about the untax. When the newspapers mention a gas tax or a carbon tax, the first response is often "Of course, it's dead on arrival," or "It's a political third rail." Mankiw puts it like this: "Republican consultants advise using the word 'tax' only if followed immediately by the word 'cut.' Democratic consultants recommend the word 'tax' be followed by 'on the rich.'"

I favor the untax because it's fair and it works, but in the real world its most important virtue is that it really isn't a tax. It's not a tax because it doesn't collect revenues for the government. Mankiw's carbon tax is similar, but he wants to implement it in place of part of the payroll tax. Not a bad idea, if you ignore politics. But taxpayers would not get a check in the mail, the government would keep the money, and Mankiw's carbon tax would be doing exactly what a real tax does now. That's "a new tax," just as Mankiw calls it in his headline.

Now, I imagine that many on the environmental side will be suspicious of a policy that is so similar to one backed by President Bush's chief economist, as well as most of the economics profession. But I would like to point out that the most famous advocate of the untax is none other than James Hansen. Hansen kick-started the global warming debate with his testimony before Congress in 1988 and is now Al Gore's science adviser. A talk he gave in June 2008 was titled "Carbon Tax and 100% Dividend—No Alligator Shoes!" "Alligator shoes" refers to the lobbyists who will try to get their hands on the untax revenues, and Hansen says our motto should be 100 percent or fight! That's his way of saying "Don't touch the untax."

Some people will want to change the untax to pay down corporate taxes, while others will want to spend it on energy programs. Both of these options change the untax into a regular old we-hate-it tax. Let me make this as simple as possible:

▶ If the government keeps the money, it's a tax.
▶ If it's a tax, you can forget it; it will never fly—especially if it's strong enough to make a difference.

As I show in Chapter 18, the untax is more fair than a tax—even a tax that is fully offset by reductions in other taxes. But that's not the point. As a true, verifiable, 100 percent untax, I think it has a good chance of becoming reality. But touch the untax revenues, and the untax vanishes in a puff of politics. The revenues belong to the American people.

~

The best energy plan fixes the problems of the energy market rather than just addressing symptoms. First, the energy market fails to price in the costs of climate change. So tax carbon and refund all the revenues on an equal-per-person basis—that's an untax. Because OPEC and oil prices cause even more problems, use the untax to stabilize the price of oil. This will help investors in alternative energy sources.

Because consumers ignore part of future energy savings, reward carmakers for fuel economy in the amount of the overlooked savings. Design this race to fuel economy so that it helps, rather than hurts, the Big Three carmakers. Finally, because the market fails to fully reward advanced research, increase government funding for research substantially.

This simple prescription, which includes no laundry list of complex subsidies and tax loopholes, will do most of what we need and far more than any previous energy policy.

Part 2

Energy-Market Realities

chapter 8

Learning from OPEC

After a decade's bonanza, the Saudis found their cartel losing its power; its soaring prices had shrunk demand.

—William Safire, January 1986

OPEC MEETS TWO OR THREE TIMES a year to set the amount of oil each of its fourteen member countries will produce. The cartel does not keep secret its market manipulations; you can find its "Crude Oil Production Allocations" right here on the Web:

www.opec.org/home/Production/productionLevels.pdf

OPEC, the Organization of Petroleum Exporting Countries, controls the world price of oil by controlling its production. Were OPEC to cut production 10 percent, the resulting shortage would send the world price of oil higher than we have ever seen. The organization doesn't do this for two reasons. First, its members find it hard to agree on which of them will cut back and by how much. They also know that the world would take one look at such high prices and begin to cut oil use, just as it did once before. Let's take a look back at this history to understand better the process of conserving oil and energy and why it frightens OPEC.

OPEC tripled the price of oil in 1974, then doubled the resulting price in 1979. By 1981, a worldwide reaction forced Saudi Arabia, OPEC's leading supplier, to cut production in order to keep the price from falling below

OPEC's target level. By the end of 1985, Saudi Arabia had cut its production 75 percent and could afford no more cuts. It abandoned the cartel rules, stole business from other cartel members, and let the price collapse. This ended a twelve-year price shock that is by far the largest experiment in energy policy ever conducted. The experiment did much harm and, quite by accident, much good as well. The results surprised people in three ways:

► The high prices triggered more conservation than most experts had thought possible.
► This conservation brought down the price of oil for eighteen years.
► High energy prices led to reductions in carbon dioxide emissions.

The importance of the carbon dioxide reduction did not become apparent until later.

High Oil Prices Drive Conservation

By 1986, "the Saudis found their cartel losing its power; its soaring prices had shrunk demand." William Safire, the well-known *New York Times* columnist and a self-described "right-winger," provides this analysis in the chapter's opening quote and goes on to make clear he's talking about conservation. Safire's remark demonstrates that in 1986, conservation was not a partisan concept. Conservation, with a little help from non-OPEC supplies of oil, defeated the mighty OPEC cartel. Conservation is the main way the world responds to high market prices. When price goes up, consumption comes down—but it takes a while for the full price effect to play out.

Market-driven conservation is a slow process—slow to get going and even slower to stop. Looking at recent high oil prices, people noticed that gasoline use was slightly higher in 2006 than in 2005, and many concluded that higher prices were not working to curb gas consumption. People thought the same in 1974, when the price of oil tripled and world oil consumption fell only 1 percent.

Market-driven conservation starts slowly because the best way to conserve is to switch to better technology. People don't buy cars and refrigerators until they need new ones, and companies take years to design new, more efficient models. It takes a while for changes in technology to pay off. But starting in 1980, with new technology in place and oil prices spiking, Figure 1 shows world oil use taking an unprecedented four-year nosedive. Figure 1 also shows that people kept conserving after the oil price collapse. In fact, changes made in 1980 are still saving us oil, otherwise the price of oil would have hit $100 a barrel years ago.

The Department of Energy (DOE) documented the unexpected size of the OPEC conservation effect back in 1980, and William Nordhaus, a respected Yale economist, discussed it in the *New York Times* that same year. Dale W.

Figure 1. OPEC Raised the Price, and the World Conserved Oil

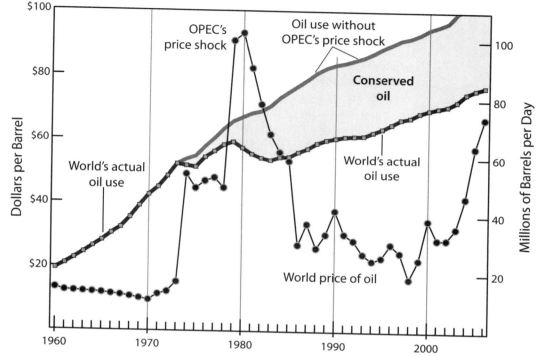

The top line is estimated world oil use without the two OPEC crises. The line that branches off below it in 1974 is actual world oil use. The difference is the amount of oil conserved because of OPEC's high prices. Notice that changes made because of OPEC—things like fuel-economy standards and better insulation—are still saving an enormous amount of oil worldwide. Oil prices are in 2007 dollars.*

Jorgenson, whom I cited in Chapter 2, and Peter J. Wilcoxen are two of the country's best applied economists. They intensively studied the impact of the oil shocks on the United States and concluded that "over the period 1972–1987 U.S. emissions of carbon dioxide were stabilized by *price-induced energy conservation* [emphasis added]." Although carbon dioxide emissions worldwide did not stop increasing, they did stop increasing in the United States—for fifteen years. And during the crisis, global emissions also increased more slowly.

The Power of Price

The power of price lies in its ability to act in a million ways at once, many unexpected. Even when price directly affects people, they don't always recognize it. For example, consumers upset with high gas prices in 1975 lobbied for

Corporate Average Fuel Economy (CAFE) standards, federal regulations that require improved fuel efficiency in vehicles. These mileage standards continue to affect car buyers to the present day, but few recognize the role of OPEC's high prices in bringing about these energy-saving measures. Many people also failed to notice that the collapse of OPEC's price caused the freeze in mileage standards from 1985 until 2007. Lawmakers have revived increases in mileage standards only because oil prices have again risen for several years running. Even the energy gurus of the physics camp, who now push for stricter standards and ignore energy prices, owe their careers to OPEC's high prices. I say this not to belittle their work, but to point out how fundamental and varied the price effect is. Price changes everything. And the whole world responded to OPEC's high prices.

As Figure 2 shows, high prices also lead to increased supply. New oil supply generally requires new wells, and these take time to develop. As you can see in the graph, it took about five years after the first major price increase for supply to increase noticeably, and it took about seven years after prices declined, until 1993, for the extra supply to evaporate. The extra non-OPEC oil supply over the years did not total up to even one extra year of oil supply measured at the 2006 level. On the other hand, conservation provided us with the equivalent of eight or more years' worth of extra oil (see Figure 1). Conservation gave us ten times more bang for the OPEC buck than increased supply. Even today, the leftover conservation measures from the 1974 to 1985 OPEC crisis are doing more for us than the extra supply did at its peak in 1985.

Did an Oil Glut Cause Prices to Fall?

The most dramatic change shown in Figures 1 and 2 is not the enormous conservation effort or the rise in non-OPEC oil production, but the rise and fall of the oil price itself. The price increased to six times its 1973 level, then plunged to less than a third of that new high. What caused these changes?

The oil embargo of 1973 and the Iranian revolution in 1979 sparked the price increases. But these two events do not explain the bulk of what happened. They only triggered OPEC's quest to increase profits by raising prices and cutting production. The upswings are just normal price gouging. But the price decrease is more puzzling.

Markets have a way of getting even. When some suppliers push the price up, the high price motivates consumers and other suppliers to take actions that push it back down. As we have just seen, OPEC's massive price hikes caused the two standard reactions—increased supply and reduced demand. Both changes happened slowly, so OPEC was able to hang on to its profits for several years.

Both increased supply and decreased demand lead toward a glut of unsold oil, which frustrates suppliers trying to sell their product. The most

Figure 2. High Prices Increase Supply but by Less than the Conservation Effect Shown in Figure 1

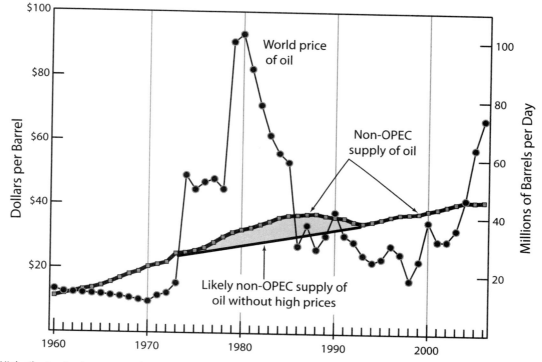

High oil prices lead to more exploration for oil and a greater supply of oil from non-OPEC countries. This is the cause of the "bump" in the non-OPEC oil supply line between 1976 and 1993. The extra supply shown here in gray is much less than the supply saved by conservation, shown in Figure 1. The two figures use the same scale to facilitate comparisons.*

effective way to sell is to cut the price, which OPEC did. But was an oil glut really why OPEC cut the price? It is important to be sure, especially if our national goal is to force such price cuts again. When OPEC cuts prices, it often gives a reason for the price reduction, such as a concern for the world economy. However, this is just part of the game. It is best to check what was actually happening when OPEC cut the price. The DOE maintains records of events in the world oil market, and this is part of the department's history of that period:

► "1982. Indications of a world oil glut lead to a rapid decline in world oil prices early in the year. OPEC appears to lose control over world oil prices.

► 1983. Oil glut takes hold. Demand for oil falls as a result of conservation, use of other fuels, and recession.

▶ 1985. OPEC loses customers to cheaper North Sea oil. More OPEC price cuts."

History confirms that an oil glut is what put pressure on OPEC's price. When demand decreases or supply increases, suppliers cannot sell a portion of their oil until the price falls.

Although most of the story is just this straightforward, an unusual event occurred when the oil price first peaked:

▶ "1981. Saudi Arabia, a member of OPEC, floods the market with inexpensive oil, forcing unprecedented price cuts by other OPEC members. In October, all thirteen OPEC members align on a compromise [lower] $32-per-barrel benchmark (in 1981 dollars)."

Why would a near monopolist flood the market? Saudi officials of the time would tell us they did so to set a lower, more reasonable price. Obviously they knew flooding the market would bring the price down, just as it did, but why did they want a lower price? Periodically, OPEC has lowered prices, and its members always make a fuss about how responsible they are being and how we all want a "stable" price.

The reality is different. The Saudis, in particular Ahmed Zaki Yamani, Saudi Arabia's oil minister from 1962 to 1986, wanted a lower price because he was afraid OPEC's extremely high price would soon bring a market response strong enough to crush that price. Yamani was right. Unfortunately for him, he could not get the other members of OPEC to lower the price to a level that was sustainable. Six years later, he was losing so much money from the oil glut caused by high prices that he started taking business away from other cartel members. This caused a complete price collapse, which disciplined the other cartel members, and the cartel is stronger for it now. Yamani, however, lost his job in the process.

OPEC's motives are simple. Its members want to make as much money as possible over the long run. This means they want the price of oil as high as possible without causing a market response strong enough to force the price back down. When OPEC overreaches, consuming nations react with strong conservation measures that push the price down again. OPEC has learned the hard way that this destroys long-term profits. Notice in Figure 2 the eighteen years of low prices OPEC suffered the last time it overreached. This time it is being more cautious, but has it been cautious enough? In a world richer than it used to be, with demand booming in developing countries, OPEC is betting it can keep the price high indefinitely.

Figure 3. Dick Cheney's Graph of OPEC-Induced U.S. Energy Conservation

This graph is Figure 8.1 from the 2001 report by Dick Cheney's National Energy Policy Development Group. It shows OPEC's enormous and enduring influence on conservation.

From 1950 through 1973, energy use is almost perfectly predicted by GDP. But starting in 1974, the first full year of the OPEC crisis, actual oil use falls increasingly behind the historical trend. The difference between the two lines is due to conservation.

By 2000, conservation is saving about 65 quadrillion Btu, and U.S. energy use is about 100 quadrillion Btu. Forty of the 100 "quads" of energy we use comes from oil, so 65 quads of conservation is far more energy than comes from the oil we use every year. This conservation is a response to OPEC's high prices from November 1973 through 1985.*

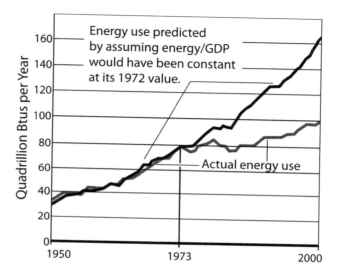

A Consumers' Cartel: Do-It-Yourself Conservation

High prices prompted the world's consumers to act as if they were part of a consumers' cartel. Consumers "cooperated" by reducing consumption to bring down OPEC's prices. Non-OPEC suppliers "cooperated" as well, though to less effect. But this pseudocooperation was just a reaction to the cost of paying tribute to OPEC, Big Oil and all other oil suppliers.

In six years, this "cooperation" brought the world's demand for oil back down to a level at which the world was safe from OPEC for another eighteen years. But the cost of this victory was enormous—as high as $4 billion per day, or $1.4 trillion dollars per year, in 1980.

Long before costs reached such extravagant levels, the world began searching for a way to cut costs. Former Secretary of State Henry Kissinger came close to finding it. In early February 1974, at a conference of thirteen oil-consuming nations, Kissinger proposed a "truly massive effort" of cooperation, according to the *New York Times*. "The United States will join with other consumer nations in a study of joint conservation policies in an effort to hold down the use of energy," reported the *Times*. By the end of the year, the International Energy Agency (IEA) had been organized, with the intention that it act as a consumers' cartel.

Had the IEA been effective, it could have prevented the second energy crisis, which doubled of prices starting in 1979. Although the IEA failed as a consumers' cartel, conditions are now more favorable for cooperation, as I will discuss in later chapters. But that early OPEC experiment can teach us how effective a cartel could be—how much conservation it takes to reduce OPEC's price of oil.

Figures 1 and 2 show that, from 1979 through 1985, a 35 percent change in *net* demand (demand reduction plus non-OPEC supply increase) caused the price of oil to drop from about $90 to about $30 a barrel (in 2007 dollars). That's a 200 percent drop if you compare it to $30, but that method exaggerates a bit. The correct method uses a compound interest type of formula, which I won't go into because the only point I wish to make is rather modest. For every 1 percent cut in world demand for oil, we should be rewarded, on average, with more—perhaps much more—than a 1 percent drop in the price of oil.

As we will see later, this is consistent with estimates that the IEA uses, the DOE uses, and other economic models also use. Cutting demand has a powerful effect on price.

~

In the chapter's opening quote, William Safire explains that OPEC lost power because "its soaring prices had shrunk demand." In the same 1986 column, he goes on to make a recommendation: "What we should do to help oil prices continue moving down to the mid-teens, and stay there, is no secret: … impose a $12-a-barrel oil import fee." (The fee would be $20 in 2007 dollars.) He points out that one virtue of the import fee would be to "encourage the continued conservation of fuel by the U.S. consumer."

High energy prices have proved themselves as the most effective tool for achieving our twin goals of climate stability and energy security. When price rises, the demand side of the market responds more quickly and more vigorously than the supply side—and its response lasts longer. Put simply, conservation is about ten times more potent than supply increases.

OPEC's great energy experiment proved it could raise prices—for a while. But we should learn a different lesson. High prices can stimulate enough conservation to bring world oil prices back down. Since that experiment cost us around a trillion dollars, we should learn what it had to teach us. It looks like we'll be needing it again.

chapter 9

The World Oil Market versus Energy Independence

Dependence on oil creates national security issues. There's too many people who have got oil that may not like us.

—George W. Bush, 2007

THE WORLD OIL MARKET CONTROLS the price you pay for gas at your neighborhood gas station. Taxes, gas station profits, and oil-refinery profits also take their toll, but when you see the price of gas go up twenty cents in one week, that's the world oil market in action. There's no escaping it. Even if your gas station sells gasoline made from 100 percent American oil, the price goes up exactly the same amount. Even if you buy American corn ethanol, the world oil market hits you just as hard.

This spells bad news for the most popular paths to energy independence: more drilling and alternative fuels. But the world oil market treats two other paths more kindly: energy conservation and electric cars. Conservation defeated OPEC from 1986 through 2002, and conservation wins again when it comes to protecting American consumers. But only electric, or perhaps hydrogen, cars can make us fully independent. They can provide non-liquid-fuel sources of transportation energy.

Oil Tankers Make the Market

Although oil tankers are expensive to build, they move so much oil so cheaply that they add relatively little to the price of oil. Cheap transportation of oil keeps oil prices aligned around the world.

For example, the United States buys more oil from Canada than from any other country, and Canadian companies can sell oil profitably for $60 a barrel. Did this help us when the world price went to $100 per barrel in early 2008? Unfortunately, it did not. Canadian companies, like all oil companies, can sell their oil anywhere in the world and pay only a small charge for transportation. So when China or Germany is paying $100 per barrel, Canada is not going to sell oil to the United States for $60 a barrel.

The ability of oil companies to sell anywhere with only a small transportation cost means no company sells oil at much below the world price. That creates a single world oil price. Because of this, it doesn't really matter who the United States buys from. Buying from Canada is no protection at all. When a shortage occurs, the price we pay goes up just the same.

A significant supply disruption anywhere in the world causes a price shock everywhere, so a world market may seem to increase the danger. But it also reduces the height of the price shock by spreading it over the whole world. In another way, having a unified world oil market provides excellent protection. OPEC cannot harm the U.S. supply of oil without harming the whole world equally.

Even if the United States bought most of its imported oil from OPEC, cutting us off would cause us no special harm. Here's what would happen. Our oil companies would immediately offer to buy oil at a bit above the world price from any oil company in the world. Since those other companies could make money by buying at the world price and selling to us for a bit more, many would be happy to do so.

The Military and Oil Security

> American forces ... are in Iraq to prevent Iranian imperialism ... from dominating the energy supplies of the industrial democracies.*
>
> —Henry Kissinger, 2007

∼

In fiscal year 2005, the U.S. Department of Defense consumed 133 million barrels of petroleum. The U.S. Strategic Petroleum Reserve stands at 688 million barrels, enough to supply all the requirements of the Department of Defense for five years at its 2005 rate of use.

In fact, they would compete to get our business, and that would keep us from having to pay much more than the going price. For a small premium above the world price of oil, we would get all the oil we wanted, in spite of OPEC.

Put more simply, if OPEC cut 5 million barrels a day from the United States or Japan or any other country, the effect would be the same. The price of oil would rise, perhaps significantly, but the world oil market would assure that nations share the pain evenly. All countries would buy less because of the high price and not because of which countries OPEC favored or embargoed. OPEC can cause a shortage and raise the price, but it cannot effectively target any country.

So the world oil market behaves in a simple fashion in at least two ways:

▶ All countries pay basically the same price for oil.
▶ Any country can buy all it wants at approximately the world price.

These two points provide enough of an understanding of the world market to analyze the three paths to oil independence.

Three Paths to Independence

The three basic approaches to energy independence differ in the way the world oil market affects them. The approaches are

▶ *Produce more* fossil, or nonfossil, liquid fuel domestically.
▶ *Use less energy* for transportation.
▶ *Use electricity* instead of liquid fuel (electric cars).[1]

Producing more has long been the strategy that the oil industry favors, although it is not at all keen on farmers producing more ethanol. In any case, the produce-more strategy is so generally popular that the public often overlooks other strategies. Producing more, however, has a dark side that needs illuminating.

The Produce-More Strategy. We can produce more liquid fuel by converting corn to ethanol, converting soybeans to diesel, drilling for oil in Alaska's Arctic National Wildlife Refuge, or converting coal to gasoline, to name some of the more prominent possibilities. The future will bring even more options—some better, some worse. The alternatives have different costs and different pros and cons. To the oil companies, *alternative fuels* mean liquid coal, shale oil, and oil from federally restricted areas. To environmentalists, alternative fuels are renewable biofuels. From an energy security perspective, alternative fuels are all the same, even though they differ sharply in their effects on global warming.

President George W. Bush promised that we will be making 35 billion gallons of alternative fuels by 2017. So you might think we are well on our way to energy independence and are at least partly protected from the next OPEC oil shock. Unfortunately, even with that much ethanol, an oil shock would hit U.S. drivers just as hard as it would without the extra fuel. But there would be one big difference: The alternative-fuel producers would make a killing. They would sell each of those 35 billion gallons of ethanol for exactly the same price as OPEC-based gasoline and pocket the price increase as profits. That's how the world market works.

---

1. Hydrogen is an alternative energy carrier that could someday be used for transportation. But currently, car companies are much more optimistic about electricity.

What about Government Price Controls?

Producing more and using less are market-based approaches to independence. But is it possible for the government to control the domestic price of oil directly and stop price shocks that way?

In 1971, even before the OPEC crisis, President Richard Nixon imposed a wage and price freeze. The oil price controls on domestically produced oil lasted until President Jimmy Carter made this announcement on April 5, 1979: "Federal government price controls now hold down our own production, and they encourage waste and increasing dependence on foreign oil. … I've decided that phased decontrol of oil prices will begin on June 1 and continue at a fairly uniform rate over the next twenty-eight months."

Lower prices sounded good, but they discouraged conservation and increased supply. Price controls worked against the two forces that saved us from OPEC. And to the extent they work, they just produce long lines at the gas pump.

This is not just a theory. We produce almost half our gasoline domestically, and the cost of producing that gasoline doesn't change at all when OPEC raises the price of oil. But when the world oil price increases, so does the retail price of all gasoline, whether it is made from domestic oil or foreign oil. You don't find low-price gas stations selling domestic gasoline.

You won't find low-price domestic ethanol stations or low-price domestic liquid coal stations either. All liquid fuel prices move together. Nebraska has tracked the wholesale price of ethanol and gasoline since 1983, and ethanol has averaged only three cents per gallon more than gasoline.

So is the produce-more strategy just a hoax? Not quite. It helps in two ways. First, as I explained in the previous Chapter, producing more (or consuming less) helps reduce the world price of oil. Producing 35 billion gallons of ethanol could reduce the world price of oil by 2 or 3 percent. Second, it means some of our gasoline dollars that would otherwise flow to OPEC or Canada will instead flow to the American or semi-American companies that make the extra liquid fuel. I say semi-American because Archer Daniels Midland, the biggest ethanol producer, as well as the big oil companies are all multinationals.

So if terrorists blow up a Saudi oil field, alternative fuels will provide no protection for American consumers. If we are using lots of American-made alternative fuel, alternative-fuel companies will make a killing off the oil price shock by charging American consumers the world price of oil.

Robert M. Gates, secretary of defense during the second George W. Bush administration, led a scenario exercise call Oil Shockwave. In it, top former government officials took part in a series of "cabinet meetings" to discuss a hypothetical unfolding energy crisis. Here's part of the 2005 report:

> The Myth of "Foreign Oil"
> Oil is a fungible global commodity that essentially has a single world benchmark price. Therefore, a supply disruption anywhere in the world affects oil consumers everywhere in the world. U.S. exposure to world price shocks is a function of the amount of oil we

consume and is not significantly affected by the ratio of "domestic" to "imported" product. The emphasis placed on foreign oil is greatly exaggerated and provides little meaningful insight.

The Oil Shockwave exercise concludes that "U.S. exposure to world price shocks … is not significantly affected by the ratio of 'domestic' to 'imported' product." That means producing ethanol or more oil domestically will not protect us from oil price shocks. So it will not make us "energy independent." This confirms our analysis of the "produce-more" approach.

The Use-Less Strategy. Secretary of Defense Gates and the Oil Shockwave report also tell us that "exposure to world price shocks is a function of the amount of oil [domestic or imported] we consume." So the path to independence is to consume less in total. In other words, the second path—the use-less strategy—works.

The math for this approach is simple. If your car uses half as much gasoline, you are hit half as hard by an oil price shock. If your car uses the same amount of liquid fuel but a different kind, you are hit just as hard. All liquid fuels change price together. Conservation provides price protection that alternative fuels fail to provide.

Aside from price shock protection, the first two paths are about the same. But conservation—using less energy—has on other advantage over producing more liquid fuel. Replacing a gallon of gasoline with corn ethanol and conserving a gallon of gasoline both reduce oil use by about the same amount. Conserving that gallon saves slightly more oil because alternative fuels, such as ethanol, use a little gasoline in the making. So conservation reduces imports slightly more than the use of alternative fuel and lowers the world oil price slightly more.

What conservation does not do is provide windfall profits to alternative-fuel companies during an oil crisis. Conservation does, however, provide profits for companies, such as automakers, that supply the technology for conserving gasoline and oil.

The Use-Electricity Strategy. Charge your electric car's battery with electricity made from coal—as most electricity is at night, and you can power your car with coal instead of oil. This approach may not conserve energy, but it does use less liquid fuel, so it does protect us from oil price shocks. Charge your battery with electricity made from wind or solar, and you can drive without any fossil fuel. Battery technology is not quite up to this challenge, but we should be seeing some plug-in hybrids by 2010.

Conserving energy can make a huge difference, but it can never completely eliminate the use of liquid fuel. Using electricity can. Someday, the United States may become completely energy independent. Unlike liquid fuels, coal is not a good substitute for oil. So an oil price shock changes the price of coal very little. If the source of our electricity is coal with carbon capture, or wind or solar power, energy independence and climate stability goals can both be met at once.

~

Oil price shocks hurt consumers and bring riches to oil companies. This has always been true and it will remain so. That is just the way markets treat producers and consumers during a shortage.

Switching from foreign gasoline to domestic gasoline or an alternative liquid fuel means staying addicted, and it means an oil price shock still hurts consumers just as much. Switching to alternative fuel simply moves some of the price shock profits from oil companies to alternative-fuel companies. These are likely to be large multinational corporations, and some may well be oil companies producing "alternative" fuels such as liquid coal.

Conserving fuel reduces our addiction and the pain of an oil price shock. A gallon not used cannot cost us anything no matter what the price of oil.

In the previous chapter, I discussed two reasons why conservation is the best strategy: It is available more quickly than increased supply, and it saves more than increased supply can replace—about ten times more, if the past is a any indication. This chapter adds a third reason why conservation dominates: Unlike increased supply, it protects consumers from oil price shocks.

chapter 10

Corn Whiskey versus the Climate

For people in production agriculture, these soaring new sources of crop demand are pretty heady stuff. They are creating ethanol euphoria.

—Keith Collins, Chief Economist,
U.S. Department of Agriculture, 2006

THE ETHANOL THAT REPLACES GASOLINE is 200-proof corn whiskey. If it stabilized the climate, there would be no shame in letting our cars drink good whiskey. But, as with most subsidies, the corn whiskey subsidy likely has more to do with local profits than with global policy. In fact, those who profit from growing corn or refining it to ethanol have experienced, as they say in the Midwest, ethanol euphoria.

In the last decade, a controversy has raged around whether corn ethanol is green. Do its production and use in place of gasoline reduce greenhouse gas emissions and help reduce global warming? This debate has consistently ignored one factor—the world oil market. As I show in this chapter, that changes everything.

As we have seen, conservation and an increased supply of non-OPEC oil forced the world price of oil down from $90 to $30 a barrel (in 2007 dollars) in the early 1980s. We have also seen that high world oil prices stimulated a huge reduction in the demand for oil. These two dramatic effects also apply to ethanol. Increasing the world's supply of ethanol works just like increasing the world's supply of oil. It reduces the price of oil, and that price reduction increases the world's use of oil. This is not rocket economics. If something gets cheaper, people buy more of it. So the world oil market translates our good deed—replacing oil with ethanol—into more oil use by the rest of the world. Fortunately, the increased use of oil by others only cancels out about a quarter of our oil replacement. But that can tip the balance.

Subsidies and Ethanol Mileage

Before tackling the mysteries of the world market, let's take a look at ethanol as you might buy it at the local gas station. Ethanol will never save you money at the gas pump. On average it costs the same per gallon as gasoline, but you can drive only two-thirds as far—or slightly less—on a gallon of ethanol.

So it takes 1.5 gallons of ethanol to replace 1 gallon of gasoline. Or to put it another way, paying $3 a gallon for ethanol is like paying $4.50 for gasoline. But you also have to pay for the subsidies for ethanol, with your income tax. The federal subsidy is fifty cents per gallon, or seventy-five cents for a gallon and a half of ethanol. That brings us up to $5.25 to replace a gallon of $3 gas, and that doesn't count the subsidies for growing the corn. President George W. Bush set a production goal of 35 billion gallons of ethanol per year, which will replace about 23 billion gallons of gasoline at an extra cost of more than $2.25 per gallon. That's close to an extra $50 billion a year, and this goal is now law.

If we're going to spend that kind of money, it makes sense to shop around. The government should have made a list of all the energy policies we could implement and how well they work. Instead, the government barely evaluated corn ethanol before deciding to spend big bucks on it. The U.S. Department of Agriculture (USDA), whose staff knows a lot about corn subsidies but not too much about climate change and energy security, did what little evaluation was done. Not surprisingly, USDA staff looked at the wrong variable—net energy.*

What's Net Energy and Why We Don't Care

The net energy of ethanol is the energy in a gallon of ethanol minus the human-supplied energy it took to make that gallon. I say "human-supplied" because the calculations don't count the solar energy absorbed by the corn plants. The USDA found that it takes 0.73 units of input energy to make 1 unit of ethanol energy, so ethanol's net energy is 1 minus 0.73, or 0.27. So according to the USDA, the net energy balance of corn ethanol production is 27 percent positive.

Some anti-ethanol professors at Cornell University and the University of California at Berkeley say the net energy balance of ethanol is negative. But their calculations look biased to me, and I don't buy it. Others come up with a net-energy figure that's more positive than 27 percent. A brouhaha over net energy has resulted. But do we care?

Suppose we used coal to run an ethanol distillery but captured all the carbon dioxide that results from burning the coal and that we pumped the carbon dioxide deep into the ground and stored it there almost permanently. Suppose it took two units of coal energy to make one unit of ethanol energy.

This fantasy ethanol has a net energy balance of negative 100 percent—it's just terrible according to net-energy theory. But it's good for the climate because it produces zero emissions of carbon dioxide and because it replaces gasoline, which produces a high level of emissions. The ethanol itself does not add to emissions because corn plants take carbon out of the atmosphere; burning the ethanol just puts that same carbon back in the air, producing no net increase in atmospheric carbon. And remember, the energy to make the fantasy ethanol came from coal whose carbon dioxide was all captured.

Since the coal was not imported, it causes no energy security problems. So replacing gasoline, 60 percent of which we make from imported oil, with local coal and local corn is a real help for energy security and the climate—provided all the carbon dioxide is captured and stored.

So in this example, ethanol has a 100 percent negative energy balance, but it's good for the climate and for energy security. So is this ethanol good, or is it bad? And why do we have conflicting results? The trouble with net-energy analysis is that "energy" is not the problem. Energy is a good thing. Actually, it's fantastic. Almost no one wants to walk everywhere. We prefer using at least some nonhuman energy to get around. The only problem is that some energy has bad side effects. It is those side effects that matter—things like climate change and energy security concerns, not to mention pollution—not the energy itself.

So we can ignore the net-energy brouhaha. That controversy is over the wrong question. The real questions are what to do about greenhouse gases and energy imports.

Is Ethanol Green?

Does the production and use of ethanol increase or decrease total greenhouse gas emissions? That is all I mean when I use the word *green* in this chapter, although producing corn ethanol leads to numerous other environmental problems. For example, a 2007 article in the *Proceedings of the National Academy of Sciences* tells us that corn agriculture is "a major source of the nitrogen inputs leading to the 'dead zone' in the Gulf of Mexico and to nitrate, nitrite, and pesticide residues in well water."

Finding out if ethanol is green takes two steps. First, we find how much greenhouse gas emissions ethanol causes compared with the emissions caused by the same amount of gasoline. (The same amount means equal energy.) Second, we assess the impact of U.S. ethanol production on the world oil market

The Global Rebound Effect

Warning: math ahead. This box shows how I derived the value of 0.26 for the global rebound effect.

1. The IEA tells us that a 1 percent reduction, O, in oil use causes a 1.5 percent reduction, P, in the world oil price.

$$P = 1.5\,O.$$

2. Nordhaus tells us that a 1 percent reduction in oil price causes a 0.24 percent increase, F, in fuel demand:

$$F = -0.24\,P.$$

3. Combining these two gives:

$$F = -0.36\,O.$$

4. The change in oil use caused by the increase, E, in ethanol is given by:

$$O = F - E.$$

5. Substituting 3 into 4 gives:

$$O = -0.36\,O - E.$$

6. A bit of algebra gives:

$$O = -0.74\,E.$$

7. Add the increase in ethanol, E, to the global reduction in oil, O, to find the effect on total liquid fuel use:

$$F = E + O.$$
$$F = E - 0.74\,E = 0.26\,E.$$

Result: If alternative fuel, E, or conservation cuts oil demand by 1 unit, the world liquid fuel price will fall and cause a global rebound effect of 0.26 units more fuel use.

Producing a gallon of alternative fuel increases total fuel consumption by 0.26 gallons while conserving a gallon, reduces total fuel consumption, but only by 0.74 gallons.

Corn ethanol has two big greenhouse gas problems. It takes a lot of heat to distill the corn liquids into 200-proof whiskey, and that takes a lot of fossil fuel—sometimes in the form of coal. Second, corn uses a tremendous amount of nitrogen fertilizer, and producers use natural gas to make that fertilizer. Nitrogen fertilizer also triggers soil microbes to release greenhouse gases.

The authors of a report in the *Proceedings of the National Academy of Sciences* added up all greenhouse gas emissions from the production and use of both ethanol and gasoline. They concluded that, for the same amount of energy, U.S. corn ethanol causes 88 percent as much global warming as the gasoline it replaces. That's a slightly favorable result, but the researchers forgot about the world market with it's global rebound effect.*

Ethanol in the World Oil Market. People often ignore the world market because it seems as if it is just too big to affect. But the point of using ethanol is to affect *global* warming and *global* energy security. We can't have it both ways. If we count the beneficial effects that ethanol might have on the world, we must count the problematic effects as well. The effects are all small, but they add up. The effect I'm concerned with works like this:

The Global Rebound Effect

More ethanol use causes
→ *less oil to be imported,* which causes
  → *a lower world "oil" price,* which causes
    → *more liquid fuel use worldwide.*

This same rebound effect occurs when we conserve oil or pump more oil from Alaska. Consuming a gallon less of oil or producing a gallon more of domestic oil reduces imports by a gallon, just as producing a gallon of ethanol does, and the world market follows the same path. In either case, I call this the global rebound effect because cutting back on the domestic demand for oil reduces its price worldwide and causes a partial rebound in the global demand for oil. Conservation still results in a net reduction in worldwide oil use, but the reduction is less than the amount conserved. The effect operates through the oil market, but remember, the oil market is really a market for all liquid fuels.

For a more concrete look at this effect, consider this example. Suppose that replacing a gallon of gasoline with ethanol results in the world consuming 0.26 more gallons of liquid fuel—mainly oil. In other words, the strength of the global rebound effect is 26 percent.

Now I wish to discover the impact of the global rebound effect on greenhouse gas emissions. As a baseline, we'll use the figure 100 percent to refer to the amount of greenhouse gas emissions caused by producing and using a gallon of gasoline. Earlier in the chapter we learned that replacing a gallon of gasoline with ethanol reduces emissions from 100 percent to only 88 percent—a reduction of 12 percent. But the resulting global rebound effect increases emissions by 26 percent.

The net effect is a 14 percent increase in emissions worldwide. If this is correct, then ethanol is not green. Making and using ethanol increases total worldwide greenhouse gas emissions.

The 0.26 value in this example is my best estimate of the actual global rebound effect (see "The Global Rebound Effect" to learn how I came up with the figure). So my conclusion stands: U.S. corn ethanol is not green.

The 0.26 value is based on two input values. The first one is from the International Energy Agency. In its world energy model, a 10 percent reduction in net demand causes the world price of oil to fall 15 percent. This is close to what I have seen in other models, and it is certainly a modest effect compared with what we saw in the early 1980s. I discuss this value in more detail in Chapter 13.

The second input value is the increase in oil use caused by a decrease in the price of oil. I took this from a July 2007 paper by William Nordhaus, a Yale economist and a leading authority on such matters.

The Global Rebound Effect (Again)

When conservation or alternative fuel production reduces the demand for oil, this reduces the world price of oil, which causes an increase in the demand for oil equal to roughly 26 percent of the initial reduction.

The global rebound effect makes it difficult for alternative fuels to break even with respect to global warming emissions, let alone make a large difference. One promising candidate, however, is ethanol made from cellulose. This is the part of plants that we humans don't eat because, unlike cows, we each have only one stomach. Early indications are that cellulosic ethanol should reduce greenhouse gases much more—possible by 60 percent. With a 26 percent global

The Mystery of Small Changes

Most people think 1 gallon would have absolutely no effect on the world market, but a billion gallons would.

But if your 1 gallon has no effect, then my 1 gallon has no effect and together they have no effect. Keep going and let everyone in the world use 1 gallon. Then add up all six billion zeros. So 6 billion gallons have no effect. But something is wrong with this logic.

Obviously one of these gallons will have an effect, and that is what folk wisdom calls the straw that broke the camel's back. Someone realized 1,000 years ago that even though one straw cannot seem to make any difference, if you keep piling them on, one eventually will have a huge impact.

We don't know who's gallon will matter, but on average we can say every gallon has the effect that I will calculate. If this logic bothers you, just multiply all my numbers by 1 billion. That example will teach the same lesson as my calculation of the effect of 1 gallon of ethanol.

rebound effect, we would still be 34 percent ahead. However, that's not quite half as good as conservation.

~

When it comes to climate change, if you ignore the world oil market, all ways of saving oil look more promising than they actually are. Ethanol, which seems to reduce greenhouse gases 12 percent compared with gasoline, actually increases greenhouse gases by 14 percent when the effects of the world oil market are taken into accounted. Conservation of gasoline—using less—is still a winner, but it saves only 74 percent instead of 100 percent of the greenhouse gases it appears to eliminate.

chapter 11

Synfuels Again?

We have a vast, untapped oil resource right here in the West that could produce more oil than the Middle East.

—Senator Orrin Hatch, 2005

SYNFUELS ARE BACK. In 1985, President Ronald Reagan killed President Jimmy Carter's Synthetic Fuels Corporation. Twenty years later, President George W. Bush signed the Oil Shale, Tar Sands, and Other Strategic Unconventional Fuels Act of 2005. *Time* magazine defined *unconventional fuels* as "gas or oil from coal, shale and tar sands," and that's exactly what *unconventional fuels* means today. Thirty years later we are starting the synfuels process over again.

What Senator Hatch says in the chapter's opening quote is right, but the "oil resource" he mentions is shale oil, along with some oil from tar sands—100 percent synfuel. That's why he sponsored the synfuels bill that President Bush signed as Section 369 of the Energy Policy Act of 2005.

The new push for synfuels is backed by the Departments of State, Defense, and Energy, not to mention Big Oil and Big Coal. But where do synfuels fit into the big picture of climate change and energy security? In October 2007, President Bush said,

> "We have a comprehensive strategy to deal with energy security and environmental quality at the same time."

His comprehensive strategy consists of noncorn ethanol, clean coal plants, nuclear power, and efficiency standards for buildings. He also favors improved fuel-economy standards. He did not mention synfuels. He almost never does. They just wouldn't fit into a strategy billed as dealing with "energy security and environmental quality at the same time." Synfuels are a bit helpful for security but about the worst thing going for the environment.

The next thing Bush said was, "You can solve one, you can solve the other," emphasizing his promise to deal with both "at the same time." President Bush's political instincts were right on target with this one. That's what people want, and that's what will work, because "joint solutions," as I call them in Chapter 1, unite the two big energy constituencies: those for energy security and those for climate stability.

Synfuels—"unconventional fossil fuels"—are such a poor idea that Bush leaves them out of his "comprehensive strategy," and his name never appears with them on any White House Web page. So why have three government departments put their clout behind synfuels?

The Next Prize: Unconventional Fossil Fuel

First came coal, then oil, then gas. The United States led the world in oil production for nearly a century, until 1974, when the Soviet Union's production surpassed ours. Now the Middle East has about two-thirds of the remaining conventional oil. But the new fossil fuel is "unconventional"—oil shale, tar sands, and liquid coal.

Oil shale is a rock containing roughly 10 percent hydrocarbons. Heat it to about 700 degrees Fahrenheit for a month, and out come oil and natural gas. Shell Oil Company has tested a method of heating the shale in the ground with electricity and pumping out the oil and gas. It takes a lot of electricity, but it's probably cheaper and better for the environment than digging it out and cooking it aboveground, as producers have done in the past.

I consult a bit in Alberta for a client that generates electricity for a tar-sands operation. The company's ecologist explains that the tar sands he's seen are not even sticky. But like oil shale, the sands release oil when heated. The quality of this oil is poor, unlike the light quality of the shale oil that companies produce by slow heating underground. U.S. tar sands amount to only 4 percent of what we have in oil shale.

The world's supply of unconventional fuel is centered where Colorado and Utah meet Wyoming. Of the 2 trillion barrels of shale oil in the United States, the best 1.2 trillion are located in these three states. That's roughly the amount of oil the world has used since oil was discovered. The rest of the world has only about half as much shale oil as these three states.

President Carter's synfuel program went into effect as oil prices crested. At those high prices, synfuel made economic sense. But as prices fell, Exxon and the other oil companies started pulling out of the projects subsidized by Carter's Synthetic Fuels Corporation. The companies were wise to get out. Oil prices were headed down and would stay down for years to come.

Now that oil prices are back up, the questions return: Can we get our hands on that 1.2 trillion barrels of oil, and what will it cost us? The Department of Energy has posted on its Web site an article from the *Oil and Gas Journal* called "Is Oil Shale America's Answer to Peak-Oil Challenge?" The article compares the difficulty of extracting oil from shale with the difficulty of extracting oil from Alberta's Athabasca tar sands. It concludes that producers can extract about a half trillion barrels of oil from our oil shale more economically than producers are extracting oil from Alberta's tar sands. Half a trillion barrels is almost double Saudi Arabia's oil reserves.

The tar-sand oil companies are already producing over a million barrels of oil per day—over 1 percent of total world oil production—at a cost of under $40 per barrel. As the London *Times* put it, "The world's dirtiest oil is producing the highest profit per barrel for Royal Dutch Shell [Shell's parent company]." That would be $21.75 aftertax profit per barrel on tar-sands oil from Alberta in 2006 when the price of oil averaged $66 per barrel. The company made only $12.41 per barrel on its conventional oil.

Shell Oil's recent experiments with shale oil have led it to claim it can produce oil from shale for $25 per barrel. So producing that half-trillion barrels of shale oil looks feasible and profitable. If the price of oil stays even as high as $50 a barrel, producers could generate over $10 trillion in profits.*

No other energy policy is dangling a $10 trillion carrot in front of the world's largest corporations. So even though synfuels are not much in the news and even though they top the list of conflict-generating policies, my money's on synfuels. When it comes to a choice between fixing the climate and $10 trillion in profits, there's no question which way Big Oil will swing.

In Chapter 3, the chapter on peak oil, I discuss the current unconventional fuels initiative, which looks much like Jimmy Carter's synfuels initiative. The oil companies want the same expensive price guarantees and loan guarantees, and the military is again promising to buy its fuel at above-market prices.

Global Warming with Synfuels

How much energy do we get for every pound of carbon we send into the atmosphere? Fossil fuels are hydrocarbons, meaning they consist of atoms of hydrogen and carbon. The hydrogen burns to make water (H_2O), while the carbon burns to make carbon dioxide (CO_2). Natural gas contains the most hydrogen—four atoms per atom of carbon—and coal contains the least. So natural gas provides the cleanest energy—that is, it emits the least carbon per unit of energy. Coal emits the most, and oil is in between.

But there is another reason why fossil fuels vary in how much carbon dioxide they emit. Some fuels take a lot of energy to produce—for example, the energy to heat shale underground. Because this energy comes from fossil fuel, it releases carbon dioxide, but it does not contribute energy to the final fuel. The worst fuel in this regard is liquid coal. Producing and using liquid coal emits 1.8 times as much carbon dioxide as producing and using gasoline made from oil.

Fossil fuels vary in quality, and synthetic fuels vary in the energy required to make them. But they can be ranked roughly as follows:

Table 1. *Relative Carbon Intensity of Fossil Fuels*

Fossil Fuel	CO_2 Emitted per Unit of Energy
Natural gas (best)	1.0
Crude oil	1.4
U.S. coal	1.8
Liquid coal (worst)	2.5

From a climate perspective, making gasoline from synfuels is about like burning coal in your car's engine.*

And let's not forget the global rebound effect. As I explained in the last chapter, world oil demand rises when an increase in the supply of synfuels reduces the price of oil. Step-by-step, here's how it works: When the United States produces and uses an additional billion barrels per year, that goes in the plus column. But that reduces imports by a billion barrels per year, which goes in the minus column. So far—no change in world oil use. But reducing our imports increases the global oil supply, lowering the world price of oil. The lower price causes a global rebound in the demand for oil, amounting to about 260 million barrels per year. This figure is based on the 26 percent global rebound effect I discuss in Chapter 10 and represents a net increase in world fossil fuel use and the resulting carbon emissions.

The combination of global rebound effect and synfuel's high CO_2 emissions per unit of energy makes any synfuel program detrimental from a climate perspective.

Synfuels and Security

Producing liquid fuel domestically does nothing to protect American consumers from oil price shocks (see Chapter 9). When a terrorist group or the

Organization of Petroleum Exporting Countries (OPEC) raises the world price of oil by cutting off some supply, the synfuel companies—the Big Oil companies—will do just what they do with domestically produced oil today. They will raise the price to match the world price of oil.

Only the oil companies benefit from synfuel during an oil shock. But what about shortages? Shortages are the most misunderstood aspect of the oil market.

Oil shortages no longer cause long gas lines they only cause high prices. Now that Nixon's price controls are gone, along with the regional quantity controls that contributed to the gas lines back in the 1970s, gas stations won't run out of gas. They will simply raise prices so high that you will buy less gas. During a shortage, prices may be painfully high, but if you are willing and able to pay the price, you will find gas available. Synfuel will not protect you from these high prices, so it will not protect you from shortages, because shortages now cause only one bad effect—high prices.

That leaves one energy security benefit that synfuel can claim. It will, as I just noted, lower the world price of oil somewhat. This benefit is spread over all the oil consumers of the world, and it takes some revenues away from OPEC.

How Synfuels Block Cooperation

As President Bush understood, joint strategies unite those concerned with energy security and those concerned with climate stability. Unity produces strong political support. In 2003, *Time* magazine quoted Colorado's former governor Richard Lamm saying: "America's energy policy is zigzagging through history like a drunk." Having reviewed thirty years' worth of news articles on the subject, I have to agree.

During most of the last thirty years, energy security was the only concern. Now that we have an additional concern—climate change—and two energy camps fighting for conflicting policies, we can expect an even more erratic and ineffectual energy policy. But if the two camps unite behind one "strategy to deal with energy security and environmental quality at the same time," as President Bush promised in 2007, we could finally have an energy policy that works—and meets both challenges at once.

The one benefit of producing synfuels is that it lowers the world price of oil, which helps consumers. Could this also foster international cooperation in the form of a consumers' cartel? The opposite is more likely. With the synfuel industry reducing the price of oil, there will be less incentive for other nations to conserve and more incentive to just take advantage of the lower prices. To foster cooperation, we need to offer a trade. If you conserve oil, we will conserve oil, and each of us will benefit from the other's effort. Lowering

Is the Global Rebound Effect Actually 100 Percent?

Usually, people ignore the global rebound effect, and count it as zero. But some ecologists and physicists say it is 100 percent. They come to this conclusion from a physical, not an economic, perspective. Surely, they say, the world will eventually burn up all the conventional oil. So in the long run, every gallon of synfuel we use adds a full gallon to all the conventional oil we'll eventually use. This logic puts the global rebound effect from synfuel production at 100 percent instead of the 26 percent that I estimate in Chapter 10. Which argument is right?

It depends on how the fossil age ends. The physical theory assumes that conventional oil use ends when the world physically runs out of oil and that synfuel use ends on some future "stop date," long before the world runs out of synfuel. If all this is true, then using more synfuel now will not reduce the total amount of conventional oil we eventually use. It just increases the amount of total liquid fuel we use in the long run. If this story is right, then the global rebound effect is 100 percent—that is, every gallon of synfuel we use increases the total amount of liquid fuel we eventually use by a gallon.

Economists tell a different story. The fossil age ends when a cheaper source of energy becomes available. Only then will the world switch away from fossil fuel. This means the total oil used will not be determined by how much is in the ground, but by the date at which alternative fuel becomes cheaper to produce than the remaining oil. The higher the tax on carbon, the sooner that date will arrive. The slower we use oil, the less we will use before the day that pumping oil comes to an end.

Both stories contain a grain of truth, because as the physical supply of oil dwindles, the cost of producing it rises. This means the correct answer is in between the 26 percent I calculated in Chapter 9 and the 100 percent claimed by the physics camp.

If the world runs out of oil very suddenly, the answer may be closer to 100 percent. This would mean every barrel of synfuel produced adds to carbon emissions and replaces no oil at all in the long run.

If, as I believe is more likely, the end of oil is determined by the timing of the alternative-fuel technology, producing synfuel now will reduce the total oil used before the switch to alternative fuel. In this case, the global rebound effect would be only a little higher than 26 percent.

the world oil price with synfuels means oil-dependent countries don't need to cooperate by using less. Put simply, increasing the supply of heroin and driving down its price do not encourage addicts to cooperate with each other to kick the habit—quite the contrary.

Without synfuel, the oil price benefits of a Kyoto-style consumers' cartel will attract oil-dependent countries such as China, the United States, Japan, India, and Germany. The higher oil prices of a world without much synfuel will be an immense help in getting these countries to sign a strong international agreement that reduces the world price of oil in an environmentally friendly way. Such help is particularly important for gaining the cooperation of the Big Two—the United States and China. Also, oil companies hooked on synfuel profits will fight hard to keep us from kicking our habit and reducing carbon emissions.

Joint Solutions

Fortunately, just as President Bush claimed, "You can solve one, you can solve the other," and there is a "comprehensive strategy to deal with energy security and environmental quality at the same time." His policies are all good "joint solutions," partial as they are.

In fact, OPEC's incredibly effective policy to crush itself by raising the price of oil very high was a "joint solution." It cut the world price of oil to one-third and eventually less while it curbed carbon dioxide emissions by encouraging conservation. The Core National Energy Plan, which I recommend in Chapter 7, does just this with its carbon untax, except it does it more broadly and effectively by targeting all fossil fuels and not just oil.

Joint solutions fall into two broad classes: conservation and low-emission energy sources. Two of Bush's joint solutions—efficiency standards for buildings and fuel-economy standards—are types of conservation. The other three—non-corn ethanol, clean coal, and nuclear power—are low-emission energy sources. However, these are just a few of the possible solutions in each category.

~

Dirty alternative fuels harm the climate, do little for energy security, and tend to derail international cooperation on energy policy. Dirty fuels include corn ethanol (when made with present production techniques) and all synfuels.

I am not opposed to synfuels per se, only to yet more subsidies for fossil fuel—and for types of fossil fuel that are even worse than oil when it comes to climate change. OPEC raises world prices, which makes us pay more to every oil company. In 2007, this OPEC subsidy to Big Oil increased their profits by tens of billions of dollars.

Government committees brought into existence by Senator Hatch's synfuel bill are now requesting subsidies for the world's richest corporations to help them exploit a national resource, oil shale, worth over $10 trillion in profits. These subsidies increase the risk of climate change while they help oil companies make even greater profits. And the companies will make their greatest profits on synfuel—just as on conventional oil—when OPEC raises prices or when terrorists strike the world's oil supply. That should be subsidy enough.

chapter 12

China, Coal, and Carbon Capture

The Department of Energy ... will embark upon a $1 billion initiative to design, build and operate the first coal-fired, emissions-free power plant—FutureGen.

—Secretary of Energy Spencer Abraham, 2003

The thing went south.

—Deputy Secretary of Energy Clay Sell, 2008

FUTUREGEN IS HISTORY. Secretary of Energy Samuel W. Bodman pulled the plug on "the thing," as his deputy called it, in January 2008. Five years earlier, Secretary of Energy Spencer Abraham had announced FutureGen would be "one of the boldest steps our nation takes toward a pollution-free energy future." He was talking about the world's first clean coal-fired power plant.

President George W. Bush touted the project for five years as big spending for clean coal—a cornerstone of his comprehensive energy strategy. "We're developing clean coal technology. We're spending over $2 billion in a ten-year period," he said in 2006. In fact, the Department of Energy (DOE) canceled FutureGen after five years, having spent only $40 million—2 percent of $2 billion. That's what the government spends on the military every forty-two minutes.*

Coal-fired power generation is the largest, fastest-growing contributor to global warming. The DOE is restarting the clean-coal project on a different track—no demonstration plant this time—but five years is a lot to lose in this race against carbon emissions. Also the new track does not include hydrogen production, so the DOE-subsidized plants will not have "zero emissions" as previously advertised. They will cut carbon dioxide (CO_2) emissions by only about 40 percent.

China is at the center of the coal problem. It built one large coal plant nearly every other day in 2006. These plants will run until at least 2046. Between now and 2030, China will build more new electric power plants than the United States now has, and most of them will be coal fired. In 2007, China passed the United States as the most prolific emitter of CO_2. India is behind both but is following a similar path; by 2050, India is projected to have a larger population than China.

Here in the United States, the DOE predicts, coal-produced electricity will grow eight times more slowly between 2010 and 2030 than it will in China, but thirty times faster than electricity from renewable energy sources.

Although the coal problem is difficult, one somewhat new technology holds promise. Producers can capture CO_2 from power plants, pump it underground, and store it there almost permanently. No one has yet done exactly this. But commercial operations have tested all key parts of the system, and one old plant we will meet shortly has come surprisingly close to the FutureGen goal.

China: Villain or Hero?

Between 1990 and 2004, China's CO_2 emissions—mainly from coal—more than doubled, an increase of 110 percent, according to the DOE. In the same period, U.S. emissions grew only 19 percent. In this respect, China set the record as the worst of all the countries and regions the DOE tracks.

But wait. President Bush's Global Climate Change Initiative, announced on Valentine's Day 2002, is a promise to reduce U.S. CO_2 *intensity* by 18 percent in ten years. Before we condemn China as the worst offender, let us first rate China by intensity, Bush's scoring method. Carbon dioxide intensity is CO_2 emissions divided by gross domestic product (GDP).

$$CO_2 \text{ intensity } = CO_2 / GDP$$

Over that same time period, 1990 to 2004, China reduced its CO_2 intensity by 65 percent, according to the DOE. That's the best record among all the countries and regions tracked by the DOE. China was the fastest-growing producer of CO_2 but showed the most improvement in CO_2 intensity.

During that same fourteen-year period, the United States reduced its CO_2 intensity by only 40 percent. The changes in both the United States and China occurred at a time when neither country had an energy policy to speak of. How did this happen? Two factors can improve CO_2 intensity: Emissions can fall, or the economy (GDP) can grow. One helps the climate, and the other does not. So intensity does not tell us much about whether the climate is getting into more trouble or not. The country causing the most trouble for the climate might, and actually did, get the best intensity score.

Checking both China's CO_2 emissions and China's CO_2 intensity helps answer the question of whether China is a hero or a villain. It's really neither. China is growing fast, and growth leads quite naturally to more emissions. Growth is good, especially in a poor country like China, so China's growth is no reason for criticism. But growth is a problem for the global environment unless the world spends a sufficient portion of its increased wealth on curbing pollution. As we've seen, that necessary portion of income—about 1 or 2 percent—is eminently affordable.

Carbon Capture and Storage

If you've downloaded Google Earth, which is free online, you can take a virtual flight over the only operating synfuel plant in the United States. To do so, copy the following address into the Fly To box and click on the little magnifying glass:

47°21'37.62"N, 101°50'19.67"W

The plant exists because of the Carter/Reagan synfuel program.

The production of synfuels emits much CO_2, as I explained in the previous chapter, but since 2000, the Great Plains Synfuels Plant has come close to solving that problem. It is now the largest example of carbon capture and storage in the world. The North Dakota plant compresses most of its CO_2 under 5,000 pounds of pressure and pumps it through a 2-foot diameter pipe for 205 miles to Weyburn, Saskatchewan. There the CO_2 is injected almost a mile underground to help breathe new life into an old oil field by thinning out whatever thick oil is left so producers can pump it out. And there the CO_2 will stay for thousands of years. Investigators are closely monitoring the oil field for geological leaks and so far have not detected any.

But this isn't quite a "clean-coal" plant. The natural gas it makes from coal contains carbon, and when it is burned it releases CO_2. The whole process is no better, from a global warming perspective, than using natural gas. But it could be, if someone changed things around a little.

In fact, if the creators of the Great Plains Synfuels Plant had built it a little differently, it would have been the world's first clean-coal electric generator. Such a plant would make hydrogen instead of natural gas. To oversimplify the process, coal, which is carbon (C), and water (H_2O) are turned into hydrogen (H_2) and carbon dioxide (CO_2). This moves the carbon's energy to the hydrogen, a clean fuel.

In other words, a synfuel plant could gasify coal into hydrogen instead of into natural gas. Power companies would then use the clean hydrogen fuel

MAC'S IDEA

The Great Plains Synfuels Plant

Planning for the Great Plains Synfuels Plant actually started in 1972, before the OPEC crisis, but President Jimmy Carter's synfuel bill, passed in June 1980, played a crucial role in pushing the plans ahead.

The next year, Ronald Reagan had won the presidency, and his secretary of energy, James B. Edwards, backed a loan guarantee for construction of the plant in North Dakota. Reagan's budget director, David Stockman, and Edward E. Noble, the chairman of the federal Synthetic Fuels Corporation, both opposed it. Reagan settled the dispute in favor of a construction loan guarantee for up to $2 billion, and the plant was built.

It began operation in 1984 and lost a lot of money while oil prices were low, but for several years now it's been in the black. It uses 6 million tons of coal each year to produce 54 billion cubic feet of synthetic natural gas, which it sells to businesses and residents of North Dakota. It also produces fertilizers, solvents, and CO_2.

to generate electricity. The CO_2 produced as a by-product of making the hydrogen would be pumped underground exactly as the companies in North Dakota and Saskatchewan are now doing. A power producer can burn the hydrogen in a standard gas turbine—basically a jet engine—just as today's most efficient electricity plants burn natural gas in turbines.

Gasifying the coal has another advantage. It makes removing other pollutants, such as mercury, cheaper and more effective than in a conventional coal-fired power plant.

Once the CO_2 is captured and compressed, the cost of transporting it is only about $1.50 per hundred miles per ton of CO_2. Typically it is pumped 3,000 feet below ground and trapped for thousands of years, dissolved, for example, in brackish water. It might seem like the CO_2 would leak out, but remember that the natural gas you use to cook with was trapped underground for more than a hundred million years. Geologists believe there is likely to be plenty of room underground for all the CO_2 we need to store, but we need more research concerning storage locations.

You might think the tricky part of clean coal is learning to store the CO_2 underground, but oil companies have been doing this for thirty years. Just as the Great Plains Synfuels Plant does, oil producers have pumped CO_2 down into old oil fields, not to get rid of the CO_2 but as a form of "enhanced oil recovery." In fact oil-recovery projects usually make extra CO_2 just for this purpose. The CO_2 dissolves in the remaining thick oil, making it thin enough to pump out. In the process, a lot of the CO_2 becomes trapped. Producers can recycle what does come out with the oil, pumping it back in again, just as the operators of the Saskatchewan oil field do with Great Plains' CO_2.

So the benefits of clean-coal power plants are twofold: They can capture and store 90 percent or more of the CO_2. And they remove other pollutants, such as mercury, more cheaply and more completely. The disadvantage is cost. At present, that cost is uncertain, which is why we need a demonstration plant. A typical estimate is that it would raise the cost of coal-fired electricity by about three cents a kilowatt-hour. Compare that with a national average retail price of about ten cents per kilowatt-hour. Of course, it will take many years to make the switch, and as clean coal technology improves, the costs may come down.

Solving the Coal Problem

Coal's CO_2 problem can be partly solved in three ways: by conserving electricity, by carbon capture, and by using alternative energy sources instead. Alternative generation includes wind, nuclear, and solar. The carbon untax I propose in Chapter 7 would encourage all these approaches equitably. The untax would raise the price of using dirty coal, making the other approaches more competitive. Since all would be favored equally, the market would choose the cheapest approach.

Initially, the most cost-effective approach would almost certainly be conservation, though the beauty of the untax is that we don't need to know which alternative is cheapest. The market will tell us. But history suggests that the largest, quickest response would be from households, businesses, and industry, as they all spend a little more on such things a new LED lighting so they can spend a little less on electricity. This could greatly slow down the need for new coal plants, and by the time we need to start building power plants again we might have clean coal technology or cheaper wind or solar.

The conservation argument holds doubly true for China. Its government still subsidizes coal, which has the opposite effect of an untax on carbon. But even with a double effect—stopping the subsidy and starting the untax—China's economic growth is so great that it will keep building coal-fired generators, just more slowly. This makes research into clean coal technology all the more urgent.

The fourth policy that I propose in Chapter 7 is government-sponsored research. Because of the high financial risk involved in building the first plant based on a new technology, new clean-coal power plants should be at the top of the list in the energy research budget. An interdisciplinary team at the Massachusetts Institute of Technology (MIT) agrees.

In 2007, MIT published the team's comprehensive study, called "The Future of Coal," which recommends steps that could put clean coal on track. Because several approaches to carbon capture and storage are possible, the MIT team recommends three to five demonstration projects, one for each of the technologies. These should be commercial-scale projects, according to the researchers. They also recommend three to five projects testing CO_2 storage in various geological formations.

The team recommends a sensible, market-based approach to subsidizing these projects. The group suggests that the government pay only for CO_2 actually

> **Is Clean Coal Clean?**
>
> Actually, it's not. The process of making clean coal reduces atmospheric emission of carbon as well as contaminants such as sulfur and mercury.
>
> But mining the coal is as messy as ever, and it takes more clean coal to yield the same amount of energy. The "clean" electricity-generating plants also appear to cause more groundwater and solid-waste problems than dirty coal plants do.

stored. Companies that wish to gain this subsidy by building a demonstration project would bid on a certain subsidy rate. The bidder who offers to build the project for the lowest subsidy wins the project and the subsidy. If the project fails, it will capture and store no CO_2, and the government will owe the bidder no money. This will force bidders to be realistic in their bids, and the process will select the project with the best chance of success as well as the lowest cost to the government.

Carbon capture and storage will add to the cost of producing electricity, so the MIT team analyzed a global $25 per ton charge on CO_2 emitted. This would provide an incentive to adopt clean coal technology. If adopted in 2015 and increased by 4 percent per year, the MIT group predicts that 60 percent of coal use would be subject to carbon capture by 2050. In spite of the increased cost of the carbon tax, it predicts coal use would expand 20 percent to 60 percent.

I would only amend this approach by substituting an untax for their carbon tax—all carbon revenues should be returned to consumers to compensate them for higher electricity costs. Especially when combined with the international policies in Part 4, such an approach seems quite feasible and the coal problem much less daunting than it appears at first.

~

Coal has now passed oil as the largest source of CO_2 emissions, and it will pull further ahead in the years to come. Because coal is used mainly to produce electricity, coal-fired generation of electricity has become the number-one greenhouse gas problem. Fortunately, coal plants don't have wheels, so they stay in one place, making it possible to capture their CO_2 and store it underground permanently.

If all new coal power plants captured their carbon, that would eventually cost the United States about 0.2 percent of GDP—two-thousandths of the value of what we produce as a nation. Clean coal technology may not be the best answer, but it is almost certainly one solution to a critical problem. It deserves a high-priority, federally sponsored research program, starting immediately, and not just the three cents per person per year we have been spending.

chapter 13

Charge It to OPEC

Few things could more quickly arouse the exporters to outrage than the prospect of a tariff in the oil-importing countries, for such a levy would transfer revenues from their [OPEC's] own treasuries back to the treasuries of the consumers.

—Daniel Yergin, *The Prize*, 1991

THE OPEC CARTEL IS LEGAL. Its thirteen members, major oil exporters all, agree to production limits about twice a year and post them on www.opec.org. These limits strongly affect the price of oil, and a $10-a-barrel price increase costs Americans an extra $70 billion a year. That's $40 billion extra profit for foreign oil and $30 billion for domestic oil. Forty billion dollars is a thousand times more than President George W. Bush spent on his clean-coal program in its first five years.

The Organization of Petroleum Exporting Countries, OPEC, is legal, but isn't there something we can do about it? As the 2001 recession got rolling, a reporter asked President Bush, "OPEC is about to cut production 1 million barrels a day [to raise the price]. What is that going to do to our struggling economy?" Bush replied,

> It is very important for there to be *stability* in a marketplace. I read some comments from the OPEC ministers who said this was just a matter to make sure the market remains *stable* and predictable [emphasis added].

Of course, the OPEC ministers always say they are just "stabilizing" the price. But for some reason, they usually stabilize the price up, not down. And by the way, Mr. President, in the United States, it is illegal for a cartel to "stabilize" prices. Instead, we prefer what we call free competition.

Today, the U.S. government has no plan to challenge OPEC and apparently no serious desire to do so.[1] Some people say the oil-consuming nations just can't agree on things, so we may as well let OPEC take us to the cleaners. Others, who know that cartels are not free-market institutions, think it would be wrong for us to organize a cartel—even though the OPEC cartel is eating our lunch. Surprisingly often, liberals take this point of view.

But America was not always like this. At one time, organizing a consumers' cartel to challenge OPEC was the highest priority of the U.S. government. It was only a partial success, but we can do better.

Could a consumers' cartel really work?

This book says it can. We can fix the climate and charge it to OPEC. To back up this claim, I must show that cutting the demand for oil will bring down the world price of oil—significantly. This is not as easy as it should be, because essentially no research is being done on designing a consumers' cartel.

But the estimates I need to show the power of a cartel are, in fact, buried in many official reports, and at the end of this chapter I expose several of these to the light of day. They show that the action of a consumers' cartel would have the required impact and perhaps much more.

Economists make such numerical estimates, so it would be reassuring to balance these numbers against the opinions of experts—preferably ones with deep roots in the world oil market. For such confirming testimony I turn to OPEC itself. Of course, they argue against a consumers' cartel, but in the process they tell us just what we need to know.

Although history provides useful lessons on how to organize a consumers' cartel, this chapter cannot answer the question of whether we can do better this time around. That answer must await Part 4 of this book. That will show that global warming has fundamentally changed the political climate. In fact, the Kyoto Protocol is a weak consumers' cartel, and success with the climate will require a stronger one. But first, we need to learn something about how cartels work and the history of America's effort to form one.

What's a Consumers' Cartel?

First, let's review the more common type of cartel, a producers' cartel—say, for example, OPEC. How does OPEC work? It could work in two ways—and in

---

1. A few legal challenges have been brought against OPEC, but all have either failed in court or failed to get off the ground. At most, OPEC might be violating a World Trade Organization (WTO) rule. If so, OPEC could just quit the WTO.

the past it has. The cartel members can agree to raise the oil price, or they can agree to limit production. As OPEC found out, the two methods have exactly the same impact on the market, and that's a key to unlocking the mysteries of cartels.

If all the OPEC countries agreed to sell oil for $200 per barrel and no less, they would soon be selling a lot less oil. Let's say their sales fell from 30 million barrels per day to 20 million after some time. Now suppose instead that they make no price agreement among themselves but agree to cut production from 30 to 20 million barrels a day over the same time period. What will happen to the price of oil?

Economics teaches a surprising lesson about the connection between price and supply. If a $200 price would knock sales down to 20 million barrels, then cutting back production to 20 million barrels will send the world price up to $200. Even though OPEC makes no effort to raise the price, desperate consumers bid it up. It doesn't matter which the producers do, raise price or cut supply; it comes out the same. Price goes to $200, and production falls to 20 million barrels. So a cartel can work either way. OPEC members now agree on production quotas simply because that agreement is easier to enforce.

Consumer cartels work the same way, but in reverse. Consuming countries could agree to import 10 million barrels a day less. That would drive the price down. Or they could agree not to pay above a certain price.

The price approach seems appealing: Let's just refuse to pay OPEC's high prices. If consuming nations really did this, it would bring the world price of oil down to the consumers' target price. But first, OPEC would stop selling oil. Only after OPEC got desperate for revenue would it accept the low price.

> ## Market Power: Whose Is Stronger?
>
> A cartel needs market power, and the bigger the cartel, the more market power it has. Double the market share of a cartel and its market power quadruples. At least that's the standard economic analysis. If OPEC countries act together, they can exercise about ten times as much market power as Saudi Arabia by itself. That's why OPEC is organized—not because the countries like each other.
>
> The United States has twice as much market share on the consumer side of the market as Saudi Arabia has as a producer. The biggest four consuming nations alone have a larger share of consumption than all of OPEC has of production. In short, the consuming nations could have just as much market power as OPEC and maybe more— if they organized.

Chances are we would get desperate for oil first, so this approach is a nonstarter. Both producers' and consumers' cartels work best by controlling quantity and letting quantity control the price.

How to Run a Consumers' Cartel

In a consumers' cartel, the consuming nations agree to reduce their consumption, and price reduction follows. Several types of agreements are possible, and the United States suggested some of them in the 1970s. First, every country could cut its imports in half. But cutting imports in half is easy for a country that imports only 4 percent of its oil and extremely difficult for one that imports 100 percent. So high-import countries, particularly Germany, rejected this approach.

An alternative to a one-size-fits-all percentage quota is individual quotas. That's what OPEC does. But unlike with production, countries do not have good ways to control their consumers. So they might try but fail to meet their quotas. Or they might pretend to try but fail. The Kyoto Protocol is having just this problem. Setting an oil-consumption quantity is like setting a cap on emissions. Both limit quantities that are hard to control. Countries agree to a quantity, and they "try" to comply, but they fail. Who can say why? And no one can know if countries are trying hard enough until it's too late.

So quantity limits are the wrong way to run a consumers' cartel.

This could be confusing, because I just said that a consumers' cartel should control quantity, not the world price of oil. That still holds true. The consumers' cartel should reduce the quantity consumed, but not by setting quantity limits. And it should not try to set the world market price directly.

Instead, Henry Kissinger proposed a brilliant end run around the problem. Set a domestic "floor price." If we try to set the world price, we must struggle with OPEC and will probably lose. But OPEC cannot stop us from keeping the domestic price of oil above a floor price of our choosing. Here's how it would work: If the floor price is $11 and the world price is $10, then each country would put a $1 tax, tariff, or untax (the best solution, in my view) on imported oil. The domestic price of oil would be $11 or $1 higher than the world price. The tariff would vary to keep the domestic price at least at the floor.

This is how to run a consumers' cartel. In fact, a floor price on oil is policy number two of the Core National Energy Plan that I propose in Chapter 7. Clearly, it's not a new idea, but it's a good one.

The floor price reduces oil use, and no one needs to agree on quantities or enforce quantities or judge if a quantity-reduction plan will work in five or ten years as promised. Countries can implement a floor price immediately, in contrast to quantity reductions. And everyone can see immediately if a country has complied. Compliance is the key to success with a consumers' cartel or a Kyoto Protocol or OPEC or whatever climate agreement comes next. If the member countries cooperate, the organization works. If they cheat, it fails. Heroic goals lead to failure. Enforcement of cooperation, gentle or otherwise, leads to success.

Not surprisingly, the easiest time to implement a floor price is when the untax rate would be zero. That's when the world price is already above the floor price—for example, $110 when the floor price is $100. Of course, $100 a barrel is a good floor price only if it's high enough to cause significant import reductions. It was last time, so suppose it is this time.

So how would a cartel floor price of $100 save us money? When the world price is higher than the floor price, OPEC is the enforcer, and its members keep the profits. But such high prices will curb imports and bring the world price down to, say, $90, and the floor price would take over domestically. The government would charge an untax of $10 per barrel. So that extra $10 per barrel stays in the United States instead of going to members of OPEC. But since the domestic price remains at $100, consumers will continue to conserve, forcing the OPEC price down further. The further OPEC's price falls, the more money we keep.

Without the floor price, OPEC's price would fall for a ways, and then imports would kick up again and keep OPEC's price from falling any more and perhaps help it rise again.

The path to energy security is to defeat the OPEC cartel with a consumers' cartel, which is well served by a domestic floor price on oil agreed to by all consuming nations. In Chapter 7, I recommend this for national policy, but in Part 4, I recommend a more flexible international carbon-tax policy. But most countries would likely implement a large portion of their carbon tax as a floor price on oil, just as Kissinger's team recommended in 1974. With that in mind, let's look back at the history of oil-consuming nations struggling to defend themselves against OPEC.

Standing Up to OPEC

The United States began standing up to OPEC less than three months after the start of the 1973 oil embargo and with startling speed led the oil-consuming nations in the formation of the International Energy Agency, the IEA.

IEA: The Consumers' Countercartel. On January 10, 1974, President Richard Nixon invited Japan and the nations of Western Europe to an organizing conference. At the February conference, Secretary of State Henry Kissinger proposed that the consumer nations make a "study of joint consumer policies in an effort to hold down the use of energy." By March, the head of OPEC "accused the major oil-consuming nations of 'conspiring' to force down the market price of oil," according to the *New York Times*. By September, the *Times* reported that participants had drafted an "extraordinarily detailed" 7,000-word proposal. The article continues:

> The immediate objective is to exert downward pressure on oil prices. … Equal sharing of oil company data was a prerequisite for shaping the consumers' counter-cartel, American officials state.

"American officials" were already calling the proposed agency a "consumers' countercartel." It was a countercartel in that it was intended to return prices to the competitive level. But any organization that intentionally changes

Be Fair to OPEC?

Hurting suppliers is not the point. OPEC will still have most of the world's oil. Its members will still make hundreds of billions of dollars in windfall profits simply because they have the good fortune to sit on virtually 100 percent of the remaining cheap oil. That should be enough. Those countries don't need cartel prices too.

Exxon and the other non-OPEC suppliers will make money just as other businesses do. They have no right to expect a free ride on OPEC's monopoly prices. But unless someone catches them price gouging, we should leave them alone. They will have good years and bad years.

Consumers should be free to organize to break free of their fossil-fuel addiction if they so choose. That this will deflate monopoly prices is no reason to bail out oil companies.

Also remember that the country hurt most by OPEC was India. Why should poor third world countries suffer so OPEC's sheikhs can build palaces? A competitive oil price is more than fair to OPEC.

the market price is a cartel, so a countercartel is itself a cartel—a consumers' cartel.

In October, the *New York Times* reported: "The U.S. proposed that major industrial nations reduce oil imports by enforcing strict energy conservation measures. Kissinger and [Secretary of the Treasury William E.] Simon urged that each nation cut back by the same percentage. The British and German officials disagreed." The first cartel strategy that Kissinger proposed was for all member countries to cut their oil imports by the same percentage. But as I explained previously, this is more difficult for nations that import a higher percentage of their oil. So participants rejected this first cartel strategy. But in November, the *Times* reported:

> A "counter cartel" of the major oil-consuming countries … is now a virtual certainty.
>
> Countries that import 80 percent of the world's oil are uniting. … The oil-consuming nations now intend to undertake a long-term program of energy conservation and accelerated development of alternative energy supplies … to break the extortionate price level. … But alternative supplies will take years to develop. The immediate challenge is to limit consumption.

At this time, the oil-consuming nations understood correctly that conservation was their main weapon for the next several years and that alternative energy was their hope for the future. They were determined to fight fire with fire and "break the extortionate price level"—in other words, break the OPEC cartel. That month, the sixteen-nation IEA was established, and it continues to this day, now with twenty-seven member nations.

By the end of November 1974, the United States had developed a new approach to coordinating the cartel. Writing in the *Times* under the headline "U.S. Oil Plan: High Price Is Key," columnist Leonard Silk called it "startling news … that the United States is now founding its strategy on the $11 price." At the time, the world price of oil was about $10.

The Federal Energy Administration had concluded that by 1985, an $11 price would cut U.S. consumption by about 4 million barrels a day but that a $7 price would fail miserably. So the plan was to have all the IEA countries adopt a floor price of $11 per barrel. Each country would impose something like an oil tariff that would keep its domestic price at the floor level even when the IEA, acting as a cartel, forced down the world oil price. This would keep OPEC in check even after the consumers' cartel succeeded.

Unfortunately, when Kissinger proposed that "all the major consuming nations join the U.S. in establishing a 'common floor' for the prices," as *Time* magazine reported, many were suspicious "that the floor plan is mainly aimed at getting the rest of the industrial world to safeguard a big U.S. investment in costlier sources of energy." In fact, the United States had largely shifted its focus from conservation to synfuels.

Only in the following year, 1976, did the members of the IEA agree on a floor price, and then they agreed on $7 a barrel, the exact price that the Federal Energy Administration had analyzed in late 1974 and concluded would not work. It didn't work. The $7 floor price had no effect; OPEC kept the world price above the domestic floor price forever after.

OPEC Strikes Again

Three years later, in early 1979, when oil prices again started a rapid assent, Americans were stunned and suspicious that they were being "ripped off" (see "Accidentally Helping OPEC").

President Jimmy Carter set in motion the full decontrol of oil prices and called for a windfall-profits tax to recycle some of the oil companies' gains from decontrol. Before the world economic summit in June, administration officials disclosed that Carter would take a tough line in favor of cooperation among oil-consuming countries. In fact, according to the *New York Times,* "Administration officials" suggested a "buyers' cartel to negotiate directly with OPEC."

About this time, Germany suggested that the United States lead a consumer effort, and the Japanese did a complete about-face. Since the embargo of 1973, their policy had been to conciliate OPEC. Kiichi Miyazawa, an adviser to the Japanese prime minister at that time (although he later became prime minister himself), made these surprising comments just a week before the Tokyo summit in 1979:

> Our immediate task is to break that cartel [OPEC]. … We should not overlook the fact that we face a suppliers' cartel. The only effective way to deal with it is to form a consumers' cartel—there is no other way. … [Saudi oil minister Ahmed Zaki] Yamani is right. The West should economize on oil.

Time magazine, not quite keeping up with the changing mood, reported that "proposals for an outright buyers' cartel to control consumption, much as OPEC controls production, are thought to be too ambitious." Perhaps that would have been so, but on the first day of the summit OPEC raised the base price of its oil by 24 percent. After the summit, the *New York Times* reported:

> In a way, although nobody wants to pronounce the dread words,
> … the Tokyo agreement amounts to a consumers' cartel.

Consumers' cartel—the dread words. OPEC had the industrial world so frightened it was afraid even to talk about forming a real organization. But the industrial nations did talk tough for a few days and signed what amounts to a cartel agreement. The main purpose of the Tokyo agreement, according to the first U.S. secretary of energy, James R. Schlesinger, was to "inhibit the capacity of OPEC to raise prices" by holding down the growth of demand. That is precisely the definition of a consumers' cartel. All seven nations pledged to hold imports through 1985 to roughly their levels in 1979.

In the end, the United States and probably all the other summit nations kept their pledges, but not because of their determination. OPEC acted as the enforcer for the agreement among the consuming nations. OPEC's high prices assured compliance—and more. Never was there a better-paid enforcer.

Six months after the summit, the Organisation for Economic Co-operation and Development (OECD), representing the twenty-four leading non-Communist industrialized countries, found them plagued with double-digit inflation and economic stagnation. As a remedy, according to the *New York Times,* the OECD's economists proposed an "oil consumers' cartel." That was near the end of 1979, and that is the last time I can find any mention of government-level proposals for a consumers' cartel.

The 1986 oil price crash was disastrous for the U.S. oil industry, but the rest of the country was ecstatic. The crash caused a national debate over whether the United States should, on its own, impose a tariff on imported oil to prevent OPEC's eventual return to power. Conservatives and liberals alike supported such a tariff, and oil interests opposed it. The oil interests, with friends like George Bush senior, won the debate. In the final days of the oil price collapse, the *Wall Street Journal* reported:

> Vice President George Bush Tuesday sparked a sharp jump in world oil prices. Mr. Bush, who departs today on a trip to Saudi Arabia and three other Middle Eastern countries, said at a news conference that he would make a plea to Saudi officials for *stability* in world oil markets [emphasis added].

There's George Bush senior using that same code word—*stability*—that his son found OPEC ministers using fifteen years later. No wonder George junior believed the OPEC ministers, as I quote him explaining in the second paragraph of this chapter.

Trying to stabilize an oil price that was in free fall after twelve painful years of high oil prices got Bush senior in a peck of trouble, even with his own administration. An editorial in the *Atlanta Journal-Constitution* asked, "Will George Bush be boiled in oil?"

But Bush senior was not trying to help the Saudis. As the *Wall Street Journal* explained: "Mr. Bush, a former oil man whose political base is in Texas," said, "Hey, we must have a strong, viable domestic [oil] industry." Nothing is better for domestic oil producers than having OPEC raise the world price of oil.*

The debate over standing up to OPEC continued through 1986 and 1987 and up until Tuesday, November 3, 1988, when George H. W. Bush was elected president. The oil interests had triumphed.

OPEC's Greatest Fear

If you want to know what strategy would work against OPEC, listen to OPEC. OPEC pays close attention to what would damage its profits. Of course, when the organization finds a threat, it doesn't tell us directly what it is. Instead, OPEC looks for some reason to criticize the threatening strategy.

As Daniel Yergin explains in this chapter's opening quote, a tariff on oil imports arouses the exporters to outrage because "such a levy would transfer revenues from their own treasuries back to the treasuries of the consumers."

A tax—actually, an untax—on imported oil is exactly what I recommend, because, as Yergin said in 1991, it would transfer revenues from OPEC's treasuries to the treasuries of the consumers. (Or, in the case of an untax, it would transfer money to the wallets of the consumers.) OPEC's displeasure with this idea is a reliable sign that it's a good idea. Selling oil is a zero-sum game: What they lose, we gain.

In recent years, a new reason for an oil tax has worried OPEC—global warming. In 2007, OPEC stated that it was "concerned that many of the so-called 'green' taxes that are currently levied on oil do not specifically help the environment. Instead, they simply go into government budgets to be spent on other things."

Is this plausible? Is OPEC concerned that green taxes aren't working well enough? In fact, OPEC knows that green taxes help the environment by reducing the use of oil. What concerns OPEC is that the taxes will work. As to the revenues simply going "into government budgets," well, that would be our government's budgets instead of their governments' budgets. Perhaps that explains their "concern."

But if OPEC doesn't like green taxes, why did it sign the Kyoto Protocol? Well, besides the fact that the treaty requires its members to do absolutely nothing, they want to remain part of the international climate-change process. In another 2007 statement, OPEC stated that

> [OPEC] participates in many international meetings in order to
> remind governments and others who are debating environmental

Accidentally Helping OPEC

As Kissinger and President Gerald Ford struggled to put in place an aggressive policy to fight the OPEC cartel, they ran up against an American public that refused to believe there was an energy crisis. Congress didn't help. Democratic Senator Scoop Jackson, in a public hearing, declared, "The American people want to know if this so-called energy crisis is only a pretext." And the Democratic Congress fought for a low floor price and then for loopholes in the floor. Ford fought back by imposing a $1 tariff on imported oil. Four months later, he raised it to $2, which added 20 percent to the $10 cost of foreign oil.

Ford had planned a $3 tariff, but after months of wrangling he threw in the towel. Instead, the Democratic Congress forced an immediate 12 percent rollback in the price of "old" domestic oil, which was still under Nixon's price controls. Congress shifted energy policy into reverse.

European nations proved no stronger on cooperation, though individually they did more than the United States. In 1974, taxes accounted for about 71 percent of the price of gasoline in Paris and about 25 percent in Chicago. To this day, OPEC publishes an annual report on what a terrible idea the European gasoline taxes are—and OPEC holds up the United States as an example of how to be nice to OPEC and set low gasoline taxes.

In early 1979, *New York Times* columnist Leonard Silk wrote that many Americans were "skeptical that a shortage even exists" and suspicious that they were being "ripped off." *Time* magazine reported that 69 percent of the public still believed there was no energy crisis, but that prices were rising "merely because the oil companies want to make more money." Apparently the public believed that up until 1974, the oil companies didn't want to make more money.

policies that they must consider the needs of developing countries, especially those that rely on their income from oil.

"Those that rely on their income from oil"—would they, by any chance, be the OPEC countries? So OPEC participates in climate-change conferences to protect its "income from oil." How thoughtful. But although OPEC does remind us of the needs of developing countries, it might be a bit more accurate to characterize what it does as trying to stir up trouble. "It is unfair and unrealistic," according to OPEC, "to ask for more stringent commitments for developing countries over and above those already embraced by them in the Kyoto Protocol."

"More stringent commitments"—I suppose that would mean "any commitments at all," since developing countries currently have no commitments under the Kyoto Protocol. I can think of one excellent commitment they should make: Developing countries should commit to stopping their subsidies for fossil fuels—in other words, to stop wasting money and subsidizing global warming at the same time. Of course, the countries that top the list of oil subsidizers are the OPEC countries. But this is getting off the point.

My point is that OPEC members fear effective climate-change policy, and most of all they fear "green taxes" or green untaxes on oil. OPEC members fear these taxes because they know the taxes work—they reduce the use of oil. Even more, OPEC members fear such taxes because when oil use falls, the price of oil falls. And that's what really hurts OPEC's members. "We also need to be sure," according to OPEC, "that there will be enough demand for that oil and that we will get a reasonable price."

Now, what price would OPEC consider "reasonable"? Might that be the highest possible sustainable price? That's certainly what I would mean, were I in OPEC's shoes. OPEC is always talking about "security of demand." But its members are not concerned with a sudden demand disruption due to a terrorist attack on the United States. They are concerned that we might reduce our oil addiction over the long run. Pushers are always concerned about how to keep their users hooked. According to OPEC,

> Oil demand is also greatly affected by consuming countries' poli-
> cies. Taxation of energy products is often seen not only as a means
> of raising revenue, but also as a means of controlling demand in
> addressing environment and energy security issues.

So there you have it straight from the horse's mouth. "Oil demand is also greatly affected by consuming countries' policies." And what policy tops the list? "Taxation of energy products." Might these "energy products" be oil? Can't OPEC ever say what it means? Well, if I were them, I wouldn't either.

How Strong Would a Consumers' Cartel Be?

OPEC members' big worry is a tax on oil, but just how worried should they be? In part, that depends on how hard it is for consumers to push down oil prices. To do this, consumers must reduce oil demand worldwide, but how much good will that do? Of course, that depends on how much consumers reduce their use of oil. What we would like to know is the relationship between oil-use reduction and oil price reduction.

For example, would a 10 percent reduction in global oil use cause a 10 percent reduction in the price of oil, or would it cause a larger or smaller change? I will call the ratio of percent oil use reduction to percent oil price reduction the oil-use-change-to-oil-price-change ratio or, for short, the oil-change ratio (which I hope is more intuitive than the economists' term, the "inverse price elasticity of demand").

So what is the world's oil-change ratio? As it turns out, I can't pin it down, but it looks like a ratio of 1-to-1.5 would be a very safe bet. That means that each 1 percent reduction in oil use would cause a 1.5 percent reduction in the price of oil. So a 10 percent reduction in oil use would cause a 15 percent reduction in the price of oil. This ratio tells us how much a cartel must shrink demand to have a certain impact on the world price of oil.

Unfortunately, I cannot find a single estimate of this important ratio in the economic literature. Estimates must exist, because economic models use the ratio. I think the problem is that the estimates are basically professional judgments, and so far no economist has been willing to spotlight such an uncertain but important judgment.

This leaves only two ways to discover the value of the oil-change ratio. First, I can look to history and try to make an estimate. Second, I can look at the results of economic models to see what ratio they must be using. I will do a little of both and then make a cautious choice.

The Power of the Oil-Change Ratio

Suppose that when the world reduces demand for oil by 1 percent, the world price of oil falls 1.5 percent. How much money does that save consumers?

There's no trick to the problem; the total savings is 2.5 percent (or extremely close to that). For small changes, the two effects just add together.

This means that saving a barrel of oil saves consumers worldwide an additional 1.5 times as much money as the barrel cost, because the reduced demand lowers the world price.

That's why a consumers' cartel is so important. If I conserve 1 percent, I save 1 percent. But if all consumers conserve 1 percent, we all save 2.5 percent.

∼

Notice that this result assumes oil conservation is driven by a global climate organization using global carbon pricing as described in Part 4. That is why I can ignore the global rebound effect, which interferes with conservation and world-oil-price reductions.

The Great Energy Experiment. Before looking at the economic models, recall that OPEC conducted what I call, in Chapter 1, the great energy experiment. This "experiment" tested the value of the oil-change ratio by raising the price of oil sharply and then waiting for conservation to reduce demand. In Chapter 8, we saw that conservation caused a collapse in the world price of oil between 1981 and 1986. At that time every 1 percent drop in net demand caused far more than a 1 percent reduction in the world oil price.

The Economic Models. Three major studies, one by the Department of Energy, of the impact of the Kyoto Protocol apparently used oil-change ratios of 1-to-4 and 1-to-5 (see endnotes).

A 2007 MIT report on congressional cap-and-trade bills does not report enough information to determine a ratio. However, it does find that a strong international climate-change program could reduce oil prices from an estimated $90 in 2050 to an astoundingly low $48 a barrel.*

In the end, the oil-change ratio I have chosen to use is from the IEA. Organized by the United States in 1974 to confront OPEC, it is now the world's leading energy research institute and publishes the *World Energy Outlook* each year. This report includes both a "reference scenario"—a picture of what would happen with no new government energy policies—and an "alternative policy scenario" that assumes reduced fossil energy use. I have chosen to use the value for the oil-change ratio that the IEA used in its 2005 alternative policy scenario. This is the most conservative value I've found, and it is the most clearly explained. Here's the IEA's prediction of the change in demand for oil in the alternative policy scenario:

> Demand for oil in the Alternative Policy Scenario rises to just under 5000 million tons in 2030, 580 million tons, or 10%, lower than in the Reference Scenario.

This 10 percent decrease in the use of oil relative to the IEA's reference scenario would reduce the price of oil, and the IEA tells us:

> The oil price averages $33 per barrel in the Alternative Policy Scenario. This is $6, or 15%, lower than in the Reference Scenario, because lower demand depresses prices.

The IEA projects that a 10 percent reduction in global oil use will lead to a 15 percent reduction in the price of oil. I use this ratio throughout the book.

Each 1 percent decrease in the world's demand for oil causes a 1.5 percent decrease in the word oil price.

But one more estimate of the oil-change ratio deserves attention. It is more recent and may be the most relevant to conditions we are likely to face. The IEA's *World Energy Outlook 2007* considers a "high-growth" scenario in which demand for energy is high. Higher oil demand pushes the price of oil up and "international oil prices reach $87 per barrel in year-2006 dollars in 2030, 40% higher than in the Reference Scenario."

This is astounding, because in the high-demand scenario, global demand for oil is only 3 percent higher than in the reference scenario. So the IEA is using a 1-to-12 oil-change ratio. A 3 percent increase in demand (actually a hair more) causes a 40 percent increase in the price of oil. If the world finds itself in a high-oil-demand situation, then a demand-reduction policy—a consumers' cartel—would be extremely valuable. Even if it reduced demand only 3 percent, this would cancel out the 40 percent oil price increase.

In other words, if the oil market turns out to be tight—as it has been for a few years and promises to be in the future—a consumers' cartel could have a far more beneficial impact than predicted by a 1-to-1.5 oil-change ratio. In spite of this, I will stick with the most cautious estimate, the IEA's oil-change ratio of 1-to-1.5 from its 2005 report.

This is a long-run effect, which applies to changes in demand that last for many years. The short-run effect is stronger. As we have seen recently, small changes in supply and demand have sent the price of oil skittering up and down. Between 1998 and the start of 2008, the world's use of oil increased only 13

Update on Charging OPEC

After I wrote this chapter, picking a 1-to-1.5 oil-change ratio, the DOE published a new study in May 2008, "Analysis of Crude Oil Production in the Arctic National Wildlife Refuge."

Increased production has the same impact on world oil prices as reduced demand, so this report implicitly gives the official U.S. government estimate of the oil-change ratio.

On page 11, we find that the DOE estimates that a 1.2 percent increase in world oil supply causes a 1.94 percent reduction in the price of oil. That's an oil-change ratio of 1.62. Undoubtedly, this is a cautious estimate.

percent, but the price of oil increased roughly 700 percent. That's a short-run oil-change ratio of about 1-to-50.

～

Since Vice President George H. W. Bush flew to Saudi Arabia in 1986 to try to stanch the oil price collapse, the U.S. government has been trying to curry favor with OPEC, a policy that has never paid off except for the oil companies. Instead, as Japan's Prime Minister Kiichi Miyazawa said long ago, "A consumers' cartel—there is no other way."

During the early years of the first two OPEC crises, the United States organized two consumers' cartels, the IEA and the Tokyo agreement. These fizzled for lack of commitment, but they taught valuable lessons. The key lesson is that a consumers' cartel cannot work by negotiating prices with OPEC. Actions speak louder than words, and we must force a change in the market price by reducing demand. This is best done by taxing oil and, I hope, refunding all the tax revenues to consumers.

OPEC's fear of green taxes placed on oil provides a strong indication that a consumers' cartel would work. This is backed up by every economic model I have found that takes account of the impact of demand on the world oil price and is also backed up by the historical record of the first two OPEC crises. The weakest predicted effect is a 1.5 percent drop in price for every 1 percent drop in demand.

An effect of this size means that every dollar of oil not purchased saves that dollar and saves consumers worldwide another $1.50 in reduced oil prices. Even by itself, the United States could shift tens of billions of dollars of climate-change costs to OPEC and the other oil suppliers. But with a consumers' cartel, the world really could "charge it to OPEC"—and Exxon and BP and all the rest.

P.S. An example calculation of how the U.S. part of a global climate policy could be charged to OPEC is provided at the end of Part 4.

chapter 14

A Market-Based Carbon Tax?

Among policy wonks like me, there is a broad consensus ... we need a global carbon tax.

—Former Council of Economic Advisers Chairman
N. Gregory Mankiw, 2007

"IF ALL ECONOMISTS WERE LAID END TO END, they would not reach a conclusion." So said George Bernard Shaw, who knew that economists follow every recommendation with "On the other hand ..." President Harry S. Truman, who instituted the Council of Economic Advisers, learned this too late and was soon begging for a "one-armed economist."

The propensity of economists to waffle makes N. Gregory Mankiw's claim all the more startling: "Among policy wonks like me [economists], there is a broad consensus." Economists can't reach a conclusion, never mind a consensus. But he's right—economists have reached a consensus in favor of his conclusion that "we need a global carbon tax." (See Chapter 6 for more on Mankiw's *New York Times* op-ed.)*

Because economists favor market-based approaches, their tilt toward a tax may seem paradoxical, especially since Mankiw explicitly argues against cap-and-trade programs. These programs are all about trading, which by conventional wisdom must be more market oriented than a tax. But Mankiw, George W. Bush's one time chief economist, has impeccable market-oriented credentials. With his backing and the consensus of all those economics wonks, a carbon tax must be the most market-oriented approach possible, and so it is. To explain why, this chapter unravels some of the mysteries of carbon caps and carbon taxes.

Since I favor a carbon untax rather than a carbon tax, it may seem that I am not part of Mankiw's consensus. But an untax and a tax provide identical incentives for saving carbon, because they work the same on the tax collection end. Since the economics consensus concerns only the collection end of the tax, I consider myself part of the consensus. Economists disagree (as usual) over what to do with the revenues. I say just return them equally to all consumers—that makes it an untax. Because this chapter concerns only the collection end of a carbon tax or untax, every conclusion about carbon taxes applies equally to the friendlier carbon untax.

Politicians don't mind wasting money if that's what it takes to be popular, while economists are concerned mainly with cost-effectiveness. So when the extraordinary happens, and economists reach not just a conclusion but a consensus, it's worth listening. They are out to save you money. With Congress heading straight for the cap-and-trade programs that Mankiw warns us against, there's not much time to lose.

Future Caps

To avoid confusion, I'll tell you right off the bat that there is another type of carbon cap besides the cap-and-trade variety. Carbon caps come in two flavors, political and economic. The political kind typically caps emissions on some future date and lacks enforcement. Economists do not much analyze these future caps, as I will call them, but they deserve attention because they loom large in the public debate. Unlike future caps, the caps of cap and trade limit current emissions and are enforced with fines. Now, back to the future caps.

California initiated appliance standards, and that initiative led to federal appliance standards. California also led the way on efficient building codes and was the first state to require car companies to sell electric cars. The state tied for first in the race to open electricity markets. However, innovating is risky business. California's new climate initiative has opened doors nationally for other energy policies, but will it be a huge success like appliance standards, a fizzle like the electric-car mandate, or a disaster like California's famous experiment with electricity markets? The one that caused rolling blackouts.

On September 27, 2006, Governor Arnold Schwarzenegger signed AB 32, the Global Warming Solutions Act. The act caps California's greenhouse gas emissions in 2020 at the 1990 level. The Pew Center on Global Climate Change called it "the first enforceable state-wide program in the U.S. to cap all GHG [greenhouse gas] emissions" and noted that "comprehensive climate plans combined with enforceable GHG emissions targets provide the highest certainty of significant emissions reductions." Caps are commonly thought to provide the "highest certainty," perhaps because they sound so definite.

Future caps, being political, reflect optimism. In the case of the California bill, three cost studies seem to back up that optimism. Robert N. Stavins of Harvard University reviewed the three studies and reports that the state study found that AB 32 would save $4 billion a year and create 83,000 jobs. Another study found that AB 32 would have no net cost, and a third found a savings of $55 billion per year. Every Californian would gain about $1000 per year.

For a bill that is supposed to reduce emissions 29 percent between 2012 and 2020, such savings seem almost too good to be true. And indeed, the Stavins report concludes that "these California studies substantially underestimate the cost of meeting California's 2020 target."*

Schwarzenegger's bill is a typical future cap. The cap date, 2020, occurs fourteen years after enactment. The promised results are remarkably optimistic, and it includes no penalties for failure to comply. One item in the state's press release caught my eye:

> The bill also provides the Governor the ability to invoke a safety
> valve and suspend the emissions caps for up to one year in the case
> of an emergency or significant economic harm.*

In case of significant harm, the governor can suspend the cap. But there need be no harm at all. The law states that the threat of such harm is enough. If the governor says there's a threat, who could prove otherwise? But still, the press release assures us that the governor can only slow things down by "up to one year." Well … what the press release meant to say was for one year at a time for any number of years. That's what the law allows. The cap can be suspended forever if need be, one year at a time.

If the state is about to miss its 2020 target, trying to meet it at the last minute would surely cause harm. So legally speaking, the enforcement comes down to this: California has to meet a fairly stringent cap unless it brings a note from home—I mean, from some future governor who will happily blame the problem on Schwarzenegger.

California sometimes misses its targets. It mandated that 10 percent of all cars sold in the state would be electric starting in 2003. Not even one electric car was sold in the state that year. And California estimated it would save a lot of money on its electricity market when it fired it up on April Fools' Day 1998. Instead, the market bankrupted the state's biggest utility (which designed the market), and then the state spent $40 billion to buy electricity for delivery over the next ten years. California paid a bit more than twice what the power would have cost had the state waited five months. That $20 billion loss is how Schwarzenegger became governor in midterm.

A bulletproof excuse was crafted into the future-cap law intentionally. That's how the game is played. Feel good now; make excuses later. It's an especially good game when you get to sign the bill six weeks before you're up for reelection and nothing much has to be done until after you leave office.

In spite of all this, I think the California bill will get something done, and some of what gets done will likely be cost-effective. The bill is probably a useful step. But what bothers me is that the California approach is seen as tough-minded, providing "the highest certainty of significant emissions reductions." Is banking on a free lunch in the future with easy loopholes more certain than implementing a concrete program? Compare the approach in California with the approach in the Northeast. California promised more and, as with its electricity market, selected an "innovative" approach. The new approach, called "downstream cap and trade," flew in the face of standard economics. I use passed tense—"flew"—because that "innovation" has already been ditched. The Northeast has spent its time implementing the more ordinary but more promising Regional Greenhouse Gas Initiative. The initiative has built-in penalties that enforce its cap. It's not my favorite approach, but it's a solid design and is within the range of approaches that experts are actively debating.

The presidential capping record should serve as a warning to all future-cappers. President Nixon capped future oil imports at zero in 1980. President Ford capped future oil imports at zero in 1985. President Carter, in July 1979, capped oil imports at their 1977 level. This last example was a current cap, but like a future cap, it had no teeth. That cap held until 1997, but the evidence I present in Chapter 8 indicates that its success was due to OPEC's high prices and not to Jimmy Carter's cap. Future caps are just goals. Let us turn our attention now to the real, here-and-now caps of cap-and-trade policies. They limit pollution, and they have teeth.

What Is Cap and Trade?

Cap and trade is a policy that enforces an upper limit—a cap—on the amount of pollution emitted by a group of companies, such as all the utilities in the Northeast. That's what the Regional Greenhouse Gas Initiative does. It sets one cap for the whole group, and it issues pollution permits that add up to exactly the capped amount of total emissions. Every utility must have permits for all its emissions; otherwise it has to pay a fine.

That describes the cap. The trade part applies to the pollution permits. These can be distributed in several ways, but the key to the system is that polluters can buy and sell permits—that is, they can trade permits.

Trading cannot increase the number of permits, so the cap is secure. But polluters can buy as many as they like—if they are willing to pay the price. If one polluter buys more, another has fewer permits. Total pollution remains unchanged at the cap.

This system gives the regulator control of total pollution, which is what matters to society as a whole, while the polluters retain control over how much they cut back individually. The polluters who find it most expensive to reduce emissions buy permits and avoid expensive reductions. Those who find it cheapest to reduce emissions do so and sell their permits. As a result, the cleanup is done by those for whom it is the cheapest, which reduces the total cost of emission control.

Reducing costs is good, because consumers end up paying all the abatement costs. I don't mean to puncture anyone's belief in Santa, but no, making those big corporations clean up does not mean they are going to pay the bill. They just pass the costs on—sometimes a little more, sometimes a little less, but on average 100 percent. So when cleanup costs are reduced, consumers pay less.

> **How Trading Permits Saves Money**
>
> Say it would cost one utility $10 to cut emissions by a ton and that it would cost another utility $50 to cut emissions by a ton. So the utility with the high cleanup cost buys a 1-ton permit from the other utility for $30. This saves the buyer $20 because it doesn't have to spend $50 to cut emissions. The seller makes $20 by selling the permit for $30 and spending only $10 to clean up an extra ton. So both companies come out ahead, and that's why they "trade" permits.
>
> But notice that cleaning up a ton for $10 instead of $50 has reduced total cleanup costs. Economists have done the math to show that if companies make every profitable trade, cleanup costs are minimized. That saves consumers as much as possible.

Now here's the best part of cap and trade: Allowing a free market in permits doesn't just reduce costs; it saves as much as possible. Nothing's perfect, but by using a permit market, the companies cut costs more than under any regulatory directive.

The bottom line is that permit trading finds the cheapest way to meet the regulator's cap, and all the cost savings are passed on to consumers. At least this is the simplified economic theory. Reality is not quite as rosy. But cap and trade is better than the old approach in which regulators set each company's pollution limit—usually with a one-size-fits-all formula.

The First Consensus: Carbon Pricing Is Best

Mankiw refers to a "broad consensus" that "we need a global carbon tax." That is the second consensus. Before that consensus, there was, and there still is, an even broader consensus. Over 2,600 economists, including nine recipients of the Nobel Memorial Prize in Economic Sciences, signed a statement that concludes, "The most efficient approach to slowing climate change is through … market mechanisms, such as carbon taxes or the auction of emissions permits."

This is from the "Economists' Statement on Climate Change," which the economists circulated and signed in 1997. It states that carbon taxes and cap and trade (referred to as "the auction of emissions permits") are the top two choices for an energy policy. They are tops because they are the only two broad carbon-pricing, market-based approaches.

Notice that the carbon tax is listed first—another indication of Mankiw's consensus. However, the statement basically treats the two as twins. Why do economists see them as so similar?

They are twins because both are carbon pricing policies—that is, they work by putting a price on carbon emissions. The government requires carbon users either to pay for a permit or to pay a tax. Both raise the cost of burning carbon.

But one policy caps carbon use, and the other doesn't, so surely they are quite different. So it seems. But carbon users don't tend to care if the cap is 3 billion tons a year or 2 billion tons. Most users find those numbers incomprehensible, and the figures are not what influences how much gas you put in your tank or how much coal you buy for your power plant. In fact, with a hard cap, people know that somehow the cap will be met no matter what they do individually—so they have less reason to be concerned with the total national emissions.

The only aspect of cap and trade that matters to carbon users is the price of carbon. So a cap-and-trade policy with a $30 permit price has the same effect as a $30 carbon tax. No one cares what the government calls it. Buy a permit, pay a tax—it's all the same to consumers and businessmen.

A cap-and-trade system is just a carbon tax whose rate is set by the permit market. If permits are scarce, they will cost a lot, which means a high tax rate. If more permits are available, the tax will be low.

Since both systems are just carbon taxes, a cap reduces emissions more than a straight carbon tax when and only when the permit "tax" is higher than the straight carbon tax. So the question of strength comes down to this: Will a cap-and-trade approach help convince the public to accept a higher carbon-tax rate? Some think yes; I think no, as I explain in the next chapter.

Both policies work by raising the price of carbon, and economists favor using prices. But why do economists put these two policies above all others?

Producing and burning fossil fuel does a lot of damage that no one pays for. Coal mines pollute and destroy. Burning fossil fuel warms the globe. Both impose costs. Buying oil from OPEC wastes money. Defending the oil routes and foreign oil sources costs lives and money. The oil and coal companies pay none of these costs, so fossil fuels are grossly underpriced (except for oil when OPEC jacks up its price). Underpricing is the most famous and most important problem facing energy policy. Put another way, underpricing of fossil fuel is the central energy market failure.

The economic prescription is to fix what's wrong. Treat the cause, not the symptoms. If the price is low, raise the price. Don't make rules about who can use what quantity of coal or oil. Don't throw money at solar roofs or corn ethanol. Just raise the price and fix the problem at its root.

Either a carbon tax or a cap-and-trade policy, both of which are carbon pricing policies, fix the problem of underpricing at its root. Adjusted properly, they raise the price of carbon to what it would be if it included the costs of fossil fuel's side effects. That's why 2,600 economists signed on in favor of one policy or the other.

Consensus for a Carbon Tax

The consensus that Mankiw refers to favors a carbon tax over a cap-and-trade policy. This consensus is less broad than the consensus for carbon pricing, but broader than it appears. Many economists who advocate a cap-and-trade approach actually prefer a carbon tax. They just think the politics of the T word—*tax*—rules it out. So they push instead for their second-favorite approach, cap and trade.

Just to give an idea of the extent of the carbon-tax consensus, Martin S. Feldstein, Ronald Reagan's chief economic adviser, proposed a carbon tax back in 1992. Alan Greenspan, longtime chairman of the Federal Reserve Board, said in 2006 that he favors a tax on gasoline. Mankiw, George W. Bush's chief economist, argued for a carbon tax in 2007. Liberal *New York Times* columnist Paul Krugman came out in favor of it in 2000, and liberal economist Joseph E. Stiglitz agrees with Mankiw that it's essential for making the next Kyoto Protocol work. In 2008, economists in the Congressional Budget Office issued a report favoring a carbon tax. Among noneconomists, Al Gore is a leading proponent of a carbon tax, as is James E. Hansen, the most outspoken climate scientist.

In short, Mankiw is exactly right when he begins his op-ed by saying "In the debate over global climate change, there is a yawning gap that needs to be bridged. The gap is not between environmentalists and industrialists, or between Democrats and Republicans. It is between policy wonks [economists] and political consultants."

So why do economists prefer a tax to a cap-and-trade policy when trading seems so much more market oriented? That's been our question from the start. The answer is that required permits are a tax, and the permit price—say, $30 per ton of carbon—is the cap-and-trade tax rate. The government still controls this tax rate by how tight it sets the cap. Cap and trade is just a clumsy, complicated form of carbon tax.

Cap and trade holds a partial advantage only under special conditions, which economists think do not generally apply. In particular, cap and trade has some advantage (along with its disadvantages) if we know the optimal level for a cap better than we know the optimal carbon tax rate and if we don't have time to adjust the tax rate. In fact, we know little about either of these values, though some pretend to know one and some pretend to know the other. Even if we knew these things and didn't have time to adjust the tax rate, the complications of cap and trade still make it a questionable choice.

The permit market causes the complications. Markets must be set up, brokers paid, and trades tracked by the government. Although the government's choice of cap broadly determines permit prices, the market intervenes and causes permit prices to fluctuate unpredictably. This creates risk, particularly for capital-intensive, long-term projects, such as clean-coal plants and wind and solar generators. These risks raise already high capital costs, and consumers pay what economists call a risk premium.

To dampen permit price fluctuations, regulators and legislators will likely introduce more complications—for example, international permit trading. Trade with foreign carbon markets can reduce a permit price spike in the United States by bringing in more permits. This effectively lowers the U.S. carbon cap. But if the permits come from Eastern European countries via the European market, they may represent carbon abatement that occurred because of economic collapse. In Europe people call this buying hot air, and, indeed, it simply undermines the effectiveness of the cap-and-trade system.

The Environmental Defense Fund suggests another mechanism for limiting permit prices. The government could give permits to "farmers undertaking agricultural practices that store carbon in the soil," and the farmers could sell these at a profit. Unfortunately, it is impossible to know if carbon stored in the soil will remain for a hundred years or more. So permits given out for not plowing fields could turn out to be a mistake that affects the entire market.

Such mistakes under cap and trade have an unfortunate consequence. All the permit buyers benefit from someone cheating on the market with bogus permits. So there will be a conspiracy of silence to protect the cheaters. Permit markets invite trouble. Under a tax, only the cheaters gain, and the others are often hurt by unfair competition. Honest tax payers will not protect the cheaters.

～

Carbon caps and carbon taxes both work by making carbon more expensive. For this reason they are equally market based. Adding a government-controlled permit market to determine the tax rate does not make cap and trade any more market based, but it does create problems.

The unpredictability of the carbon tax rate under cap and trade imposes risks on business, and the costs of these risks are passed on to consumers. As the permit market expands, it becomes more susceptible to gaming, and the government has less control. Better to keep energy policy straightforward by using a carbon tax. In Chapter 15 and in Part 3, more problems with cap and trade are revealed.

chapter 15

Cap-and-Trade Politics

Virtually all allowances were handed out for free under the wildly successful sulfur dioxide trading program in the U.S.

—Nathaniel Keohane,
Director of Economic Policy and Analysis,
Environmental Defense Fund, 2008

MOST ECONOMISTS, FROM LEFT TO RIGHT, agree that a carbon tax is best. But cap and trade still dominates political discussion. The public wants their emission reductions certain and their taxes hidden, or so I've heard. Understand this saying, and you will know the secret of cap and trade.

Under the sulfur dioxide trading program the government hands out 10 million 1-ton emission permits, corresponding to about half as much sulfur as their recipients emitted before the program.[1] The government gives these permits to coal plant owners in proportion to past pollution and lets them know they can emit what they want, but without a permit they'll be fined $2,000 a ton. No one emits without a permit, so this rule caps emissions. The outcome is certain, and the tax is hidden. Didn't notice any taxes, did you?

We'll find the hidden taxes shortly, but this chapter focuses on how such taxes will play out politically when the little $2-billion-per-year sulfur cap program is scaled up to a $345-billion-per-year carbon-cap program. The sulfur tax

---

1. This was the second cap-and-trade program. The first capped CFC emissions by handing out free CFC permits, which resulted in windfall profits. A tax was then imposed partly to recapture the windfall profits.

was easy to hide, but a program that taxes a family of four $4,454 per year—the price of the carbon cap program, according to one estimate—is likely to make headlines. This is especially likely when the tax increases, say, 50 percent within a single year because of speculation in the carbon permit market.

The chief way to hide the tax revenues, thereby hiding the tax, is to give away valuable carbon emission permits for free. But the European public caught on to this, and word has spread to the United States. Hence, many current proposals call for auctioning most of the permits. Auctions raise visible revenues, so current cap-and-trade bills all have ways of dividing these up, as well as ways of handing out some free permits.

But what if all the permits were auctioned and all the revenues were refunded to consumers? That would make the bitter pill of a $4,454 tax much sweeter. And the cap would still work perfectly.

Before considering the domestic politics of caps and the possibility of refunds, let us begin with a global perspective. After all, the purpose of cap and trade is to solve the global warming problem.

Do Good Caps Make Good Neighbors?

From Barack Obama to Arnold Schwarzenegger, politicians are advocating a greenhouse gas emission cap of 80 percent below the 1990 level by 2050. I'll call it the 80-by-2050 cap. This cap is meant to limit the cumulative global temperature increase to about 2 degrees centigrade, or 3.6 degrees Fahrenheit. The Council of the European Union agreed with the target of 2 degrees centigrade as far back as 1996, though it remains highly controversial among scientists and economists.

Of course, to cap global temperatures, the world must cap global emissions, not just U.S. emissions. So a policy to cap U.S. emissions only works if the rest of the world goes along. Perhaps if the United States stops dragging its feet and firmly commits to achieving this goal, other countries will follow. By implementing the 80-by-2050 cap, the United States could lead by example. But, to succeed, the example must make sense to those we hope will follow.

In a purely mechanical way, having all countries target an 80 percent reduction seems simple. But consider the 80-by-2050 cap from China's perspective.

In 1990, the Chinese were emitting about 2.5 tons of carbon dioxide per person per year, so they need to cut 80 percent from that level. In 1990, Americans were emitting about 23.4 tons per person per year. In fact, in 1990, the United States emitted more greenhouse gas than any other country. Starting out at the highest emission level gives us the highest 2050 target of any country in the world.

Adopting a carbon cap for the next forty years tells the world we think it is a fair policy. The United States is now doing its part, so all other countries should follow us. Perhaps this is not what we intend to say, but others might easily think so. Environmentalists often say that any cap higher than 80-by-2050 will put us at great risk of disastrous climate change. The unintended implication of locking in the 80-by-2050 cap is to say to China and other countries, "If you do not cut back to a level far below our target, the earth is in danger, because you are not doing your part."

This message will not make the United States a world leader but, once again, the world's stumbling block. Leaders of China, India, and Brazil have been saying this for years, and they mean it. At Kyoto they flatly rejected cutting back to 5 percent below 1990 levels. They have been rejecting any cap at all ever since. If we want to lead, we must first listen.

Fortunately, another approach is gaining attention. It offers real hope for international cooperation, without which climate policy is simply doomed. China has already surpassed the United States in carbon emissions and is speeding ahead. The solution to this predicament is the subject of Part 4 of this book. Instead of asking developing countries to accept the unfair caps they have already rejected, it requires all countries to commit to an effort level measured by the price they put on carbon.

Finding the Taxes

We've been paying the sulfur emissions "tax" for eighteen years, and almost no one notices. Of course these charges are not called taxes; that would give the game away. The government requires expensive sulfur permits, and the coal plants pass on the permit costs to consumers. Just as with an untax, the government keeps no tax revenues, but, in this case, polluters get the "refunds," not consumers. The government hands out free sulfur permits worth about $2 billion a year mostly in proportion to past pollution.

Let's take that one step at a time so we can see the tax more clearly, and let's consider carbon permits instead of sulfur permits. They work the same. Imagine two identical power plants that both emit about a million tons of carbon dioxide per year. One gets 2 million free 1-ton permits and the other gets none. So one sells 1 million permits to the other at the market price of $30 per ton. In reality it's not so uneven, but a stark example makes the principle clearer.

Now the manager of the plant that buys the permits realizes that every megawatt-hour of power he generates costs him about $25 for coal plus a $30 permit, which is $30 more than it used to cost when he didn't need permits. So with costs up by $30 he charges $30 more for his electricity.

The other manager thinks in a surprisingly similar way: If I don't generate a megawatt-hour, I will save $25 in fuel costs and I will have one more permit to sell for $30. So generating a megawatt-hour less benefits me by $55, and generating a megawatt-hour more costs me $55. So he also sells his power for $30 more per megawatt-hour than before the cap. This is not just economic theory, it's how European utilities have turned free permits into billions of dollars of increased profits.

As a result, consumers pay for every permit both power plants need—the free ones and the purchased ones. The market price of all permits are passed through in higher electricity costs. It's exactly as if the government had taxed all the carbon dioxide and the power plants had passed on the tax for consumers to pay.

Consumers pay the tax, but they pay it to the power company, not to the government. But who gets to keep the tax? Not the power company that had to buy all its permits. It spent all its "tax receipts" buying permits. But any company that is given free permits is, in effect, given the right to keep tax receipts equal to the value of those permits.

In this example, one power company was given 2 million free permits. It sold half for $30 million, and charged consumers $30 in higher electricity prices for the other half even though it got them for free.

So, in effect, cap and trade is a carbon tax with a tax rate set by the permit market and paid by consumers to the companies that are given free permits. Those are usually the companies that polluted most in the past. This is the system that Nathaniel Keohane, of the Environmental Defense Fund, calls "wildly successful" in this chapter's opening quote. Its great success was in getting power plant owners to agree to reduce emissions. That was an important achievement, and now you can see why the power plant owners agreed.

The Politics of Risky Business

A key factor in domestic politics is the impact on business. But if the carbon permits are not free, does requiring them harm business? If they have to buy permits, won't they just pass on the costs to consumers? Yes and no. Passing costs through raises the price to consumers, and people buy less of a company's product. That's bad for business. In some cases, not all of the costs can be passed through at first. That's also bad for business.

But businesses adjust. In the ten years between 1998 and 2008, the price of oil went up 800 percent, and businesses did adjust, though quite a few suffered in the process, and many are still adjusting. A carbon cap or tax would be far milder in its effect, though it would hurt coal mines more and the auto industry less.

After some adjustment time, profits return to normal, which means all the costs of higher oil prices or carbon permits are passed on. If permit prices take a huge jump, the adjustment time is long, and the losses, though temporary, are greater. So it is best to have permit prices or a carbon tax start low and adjust upward fairly gradually.

A cap-and-trade program can seem to start gradually, but permit banking changes everything. Permit banking allows companies to save up permits for later use, and all currently proposed cap-and-trade programs allow it. A group at the Massachusetts Institute of Technology (MIT) has studied an 80-by-2050 cap with permit banking and found that even if the cap starts gradually, permits will cost $53 per ton at the beginning and $65 per ton five years later. This is much higher than prices mentioned in any bill.

Why do permits cost so much at first even though the cap starts gradually? It's because of price anticipation and speculation. If the price of carbon permits were lower—say, $20—traders would buy them up, hold them for five years, and sell them for $65. So permit banking causes speculation, and speculation causes permit costs to jump to $53 a ton the day the policy starts—even though the cap is very loose in the first few years.

This huge initial jump hurts business for the first few years. Unfortunately, permit banking is useful for technical reasons, which I won't go into, and it is here to stay. After the initial period, speculation in the permit market will make prices volatile, and this will continue to harm business.

Can a Locked-In Cap Hit a Moving Target?

The primary argument for a cap is that it guarantees we will hit our target. This claim carries some truth. But, for three reasons, a target enshrined in law may well prove not to be the correct target forty years from now.

First, governments are rarely, if ever, predictive wizards. Does history provide any examples at all of government targets set accurately forty years in advance?

Second, scientists have not reached a consensus on what the cap should be. The Intergovernmental Panel on Climate Change (IPCC) takes no position on what emission level makes us safe and must be met. If the IPCC ever does name a target, it will, as it does with all its estimates, state a range of uncertainty around that target. Currently, the scientific consensus is that the uncertainty is too great to allow even an estimate of the right target.

We know enough to be worried and to get moving. We also know we will probably move too slowly, simply because of inertia. But the lack of scientific consensus means that some think the problem will prove worse than current estimates, and some think it will prove less severe.

Until science speaks more clearly, we should act cautiously. But respect for the diversity of legitimate opinions dictates adopting policies that accommodate good news as well as bad news. This also broadens support for a strong initial policy, but it means admitting that the target will likely move.

Third, if the rest of the world does not buy into cap and trade, then trading cannot allocate emission reductions efficiently. This means money will be wasted on expensive projects in the United States, while low-cost opportunities are missed in developing countries.

In short, the science, economics, and politics of the world are far too complex to warrant locking in our path forty years in advance. Like it or not, we are shooting at a moving target. We need a policy that recognizes this and builds in flexibility. A rigid cap is not that policy.

Is Safety a Bad Thing?

Almost all cap-and-trade programs come with some form of safety valve. But one corner of the environmental camp believes that safety valves will keep us anything but safe. The Union of Concerned Scientists straightforwardly declares, "A cap-and-trade program should not include a safety valve." The Environmental Defense Fund, on a Web page titled "Why Safety Valves Are Very Dangerous," calls them "failure by design."

A safety valve limits the price of pollution permits—say, to $100 a ton—by requiring the government to offer an unlimited number of permits at that price. This effectively raises the cap, if and when the permits are selling at the safety valve price. However, when permits sell for less than the safety price, as they have always done in the European carbon market, no one buys extra permits, and the cap is secure.

Some say any safety valve would destroy a cap-and-trade program. But when high permit prices turn the valve on, every emitter is still being taxed—forced to buy permits—at this high tax rate. That means the pressure to conserve is greatest when the safety valve is in use. The safety valve does not reduce conservation pressure below what it was before the valve opened; it only limits the pressure to the maximum level deemed economically safe.

Setting a cap determines emissions but not cost, so the point of a safety valve is to provide some cost certainty. Most voters reject the view that cost is no object. Although polling data indicate a large majority of Americans agree that something should be done about global warming, that majority evaporates quickly when the polling questions include moderate costs.

John Whaley, who conducted a survey for the research and strategy firm American Environics in 2007, describes the results as follows: "Telling voters that global warming will lead to environmental disaster did not lead to increased support for action on global warming. In addition, when voters were told that specific proposals would lead to higher energy costs, support for policies to limit carbon dropped dramatically." In other words, most voters place severe limits on what they are willing to spend to meet a carbon cap. A majority are opposed to any carbon tax at all.

Even environmentalists who consider such attitudes illegitimate must recognize that they are real and powerful. Although a cap without a safety valve just might become law, if voters are surprised by high costs they can—and may well—simply change the law.

But it is also important to realize that the idea of limiting costs can be legitimate. It does not indicate an immoral or antisocial attitude. Well-meaning, intelligent people can and do believe that climate risks are uncertain and that, before we go to extremes, it makes sense to learn more. It is more than a tactical error to accuse such people of advocating "failure by design."

Safety valves generate controversy because of a clash between a majority of voters, who seek to limit costs, and others who believe that cost should be no object when it comes to the 80-by-2050 cap.

These are, in part, moral judgments. My point is that both camps should recognize the legitimacy of the other's judgment. If they do, I think there is room to resolve the controversy by considering practical political consequences.

Note that the two positions lead logically to opposite views on a safety valve. Some believe that, because of the danger of climate change, any cost is justified. They logically conclude that "no safety valve" is the best policy—at least if high costs cause no backlash. Others place a limit on what they are willing to spend. They conclude that a safety valve helps them achieve what they want—spending what it takes, but only up to a certain cost limit.

Some in the no-limit camp seem unwilling to recognize either the existence or the legitimacy of the pro-limit view. To assert that a safety valve at any level is dangerous is to assert that any attempt by me to limit my cost is dangerous. In other words, no limit that I choose could possibly be legitimate.

The first step that the no-limit camp can take toward reconciling these differences is to recognize that most people, like it or not, do have serious limits on what they are willing to spend. The second step is to realize the consequences of overrunning those limits, which could be either a weak implementation or a backlash that later undermines the cap's effectiveness.

$4,454 for a Family of Four

When it comes to cost, political discussions tend to steer clear of hard numbers. Fortunately, the MIT group evaluated the 80-by-2050 cap. They estimate that the initial cost of permits will be $345 billion per year in 2007 dollars. That comes to $4,454 for a family of four.*

Families will not purchase the permits, but the cost of the permits will be passed through to consumers in the form of higher prices for electricity, gasoline, home heating, and, indeed, every other product. The revenues from these higher costs will flow to those who receive free permits, typically coal mines and refineries, and to the government to the extent that it sells the permits in an auction. Revenues transferred to coal mines and refineries will typically end up in the hands of wealthier individuals, while revenues collected by the government will often be spent on energy-related programs.

Cap-and-trade programs, unlike carbon taxes, do not generally refund the value of free and auctioned permits in visible ways such as by reducing the payroll tax or sending a check in the mail. Because of this, consumers will perceive most of the $4,454 as real net costs. For many, the cost will be comparable in size to the income tax, with political implications that need closer attention.

One issue, which many environmentalists have raised, is that a cap-and-trade tax—just like a carbon tax—is regressive. The poor pay the highest percentage rate. Although some bills before Congress include subsidies that would help some poor pay their higher utility bills, this problem has not been adequately addressed.

Moreover, the market controls permit prices, and market prices fluctuate. In fact, studies of permit prices indicate that they fluctuate more than stock prices and almost as much as oil prices. It is not unheard of for permit prices to double in a year or two. This would double the "tax" from $4,454 to $8,580 per family of four. An event like this is likely during the first ten years of the program, and such an event—even if people expect it to be short-lived—would severely jeopardize the integrity of any cap-and-trade program. From a political perspective, I think environmentalists should be demanding safety valves to keep their programs safe from voter backlash during such speculative permit-price bubbles.

The problem I see with discussions of carbon-cap and carbon-tax designs is that they do not confront the magnitude of the required incentives. Environmentalists and politicians have ignored numbers like $4,454 for a family of four. Without looking at dollar values, politicians make plans to spend the carbon permit revenues on a myriad of pet projects and payoffs to businesses to gain their buy-in.

This may help get legislation passed, but in the long run it will prove catastrophic. In the long run, if cap and trade is a tax, people will see it as a tax. Any tax of this magnitude is vulnerable, especially when it fluctuates dramatically from year to year.

This raises the fundamental question of the carbon pricing approach. Does a cap-and-trade policy or a carbon tax have to be this expensive? Shouldn't energy policy be far cheaper?

Is It Cheap or Expensive?

The MIT study found that an 80-by-2050 cap will cost $345 billion per year right at the start—over 2 percent of gross domestic product—and go up from there. But in Chapter 2, I said a start-up climate program should cost much less. Only an advanced program that cuts emissions dramatically should get into the 2 percent range. What's going on?

The permit cost is not the *net* cost to America as a whole.

Spend a dollar on a permit, and some other American gets that dollar. Suppose the government auctions all the permits and gives all the proceeds back to consumers. Now it doesn't cost a family of four $4,454; all the visible net costs vanish. The total refund equals the total paid. The only net costs are the hidden costs of reducing carbon use—which I will discuss shortly. In spite of the 100 percent refund, the cap works just as well, because the government still limits the number of permits, and the rule is the same—no emissions without permits. So a cap-and-trade system of any intensity could appear to run for free. Moreover, an equal-per-person refund completely solves the problem of the tax harming the poor. As I discuss in Chapter 18, it would actually help the poor.

Environmentalists are missing this incredibly good news. They could have a cap-and-trade program that refunds all the extra energy costs, and it would work just as well. But there's a little bad news too. Saving all this carbon is still not free, even when permit revenues are fully refunded.

But the permit costs themselves are not the costs of saving carbon. The actual costs are all hidden. The MIT study also estimates the actual cost and comes up with about $10 billion, rather less than the $345 billion permit costs. That's only at the start, but for most of the forty years actual costs are considerably less than the permit costs.

However, if the government auctions the permits and uses all the revenue to help businesses adjust, to pay for research, and to subsidize alternative energy, then much of the $345 billion permit cost will become actual costs for less wealthy consumers. This comes on top of the hidden costs, which are what actually reduce emissions.

From a political point of view, a cheap carbon cap is one that does not transfer much wealth from consumers to special interests. Unfortunately, cap and trade with banking of permits, as I discussed previously, is going to hurt business at the start and continue to hurt business as permit prices fluctuate. Consequently, business will vigorously demand compensation—and by the looks of the bills before Congress, they will get it. This makes cap and trade more expensive for consumers than a carbon tax.

Should Cap and Trade Fund Alternative Energy?

If the goal is to reduce carbon emissions, shouldn't we spend the $345 billion a year on stimulating new energy technology? That would mean auctioning all the permits and devoting the proceeds to alternative energy.

But the whole point of a carbon cap or carbon tax is that a carbon pricing policy is the cheapest policy for reducing carbon emissions. The table below shows the initial years of a carbon cap program under two different revenue assumptions: Either permit revenues are refunded to consumers, or they are spent on government-picked energy projects.

Table 1. *Initial Years under a Cap-and-Trade Program Used to Subsidize Energy Technology*

Goal	Net Cost	Result
Correct the underpricing of carbon	$10 billion	26% reduction
Subsidize energy technology	$355 billion	Who knows?

Values are from the MIT group's analysis of a cap that declines in a straight line until it reaches 80 percent below 1990 carbon emissions in 2050.

The column labeled *Net Cost* shows the hidden costs of adapting to lower carbon emissions plus expenditures on energy technology. With subsidized technology, this is no longer a cap-and-trade program; it's a huge subsidy program hidden under a cap-and-trade fig leaf. Though small and cheap, the fig leaf may do more good than the subsidies. If more emission reductions are needed, we should make the cap stronger rather than dumping $355 billion into subsidies.

～

Carbon caps impose large and unpredictable taxes that make such policies politically vulnerable. In the long run, they provide far less control than people claim for them, and as I show in Chapter 23 they provide an extraordinarily poor path to international cooperation.

It is better to minimize the real costs of carbon pricing by returning the incentive revenues to consumers. Once this is done, real costs will be surprisingly low. With a full refund, a low-cost carbon pricing policy will be more palatable and more secure. In the next chapter, I explain the nature of the real, but hidden, costs of carbon abatement.

The Psychology of Caps

Let's not forget psychology and moral values. At first blush, caps look psychologically attractive. They seem to say, "Do the right thing, no matter what the cost." But caps are wolves in sheep's clothing.

Their most obvious deception is a false sense of security. They seem to be binding. But let's look ahead. Typically, caps and the emissions they allow decline by the same number of tons each year. At first, cheap opportunities to decrease emissions are available. But each year, the cheapest options are taken, and eventually only the most expensive remain.

On the way to 2050, the cost will likely become higher than anticipated. Then we will see that caps are not at all binding. They are just as easy to change as to install. Pass another bill, the cap goes up and the costs come down.

But a more sinister aspect of cap psychology lies in the control it takes away from individuals. Caps corrode the ethos of the environmental and energy-independence communities.

Hundreds of books, Web sites, and groups and millions of individuals now promote ways we can each help save fossil fuel and reduce greenhouse gases. Even hard-nosed people like James Woolsey, a former head of the CIA and leading neoconservative, drives a small car to help fight terrorism. It's not quite rational economics, but it makes sense morally and psychologically. Caps will undermine the moral and psychological rationale for such behavior.

Suppose we have a cap-and-trade system, and you buy a small hybrid car—a little smaller and more expensive than you would like. But you want to help knock down OPEC prices and help the climate.

What have you accomplished? Under a cap, exactly nothing. The cap will be met no matter what you do. When you use less carbon, someone else automatically gets to use more. This works indirectly through the trading of permits, but it does work. More bluntly, after the permits are reshuffled, your squeezing into a small, efficient car just allows someone else to drive a big gas-guzzler. It doesn't help the climate or security one bit. The guys driving the guzzlers will be waving to the Prius owners and saying thanks—or maybe just laughing.

That's what caps do. They take all control of conservation and emissions away from individuals and small groups and give it to the authority that sets the cap. Everyone else can go home. An untax doesn't do that. When you save energy, it still matters.

Part 3

Core National Policies

chapter 16

An Untax on Carbon

We suggest a tax on carbon dioxide in which all the proceeds collected by the government would be returned to Americans each year.

—Keith Crane and James Bartis, *Washington Post*, 2007

"THERE IS A BROAD CONSENSUS in favor of a carbon tax everywhere except on Capitol Hill, where the 'T word' is anathema." So says the conservative American Enterprise Institute. The conflict between the antitax politics and the consensus creates a tension at the heart of energy policy. Capitol Hill politicians have blocked the world's best energy policy with antitax slogans.*

A carbon untax breaks the deadlock by dividing the carbon tax into two steps and fixing the expensive step. The first step of a carbon tax collects the money, and the second step gives it to the government. The first step, collecting the money, makes the carbon tax work and is the reason for the broad consensus. Collecting the carbon charge discourages fossil-fuel use. The untax does this, but it replaces the second step, "give it to the government," with "give it back." That's so different that I cannot call the untax a tax. The whole point of a tax is to collect money for the government.

The simplicity of the untax hides a number of puzzling subtleties. If consumers pay all the costs and receive all the refunds, why does it work? If it refunds 100 percent of what it collects, isn't it free? If it's free, how can it possibly be a powerful method of moving society away from fossil fuels? And if it has hidden costs, won't it be unfair to the poor? I will explain the basic workings of the carbon untax and then consider these mysteries one by one, though I leave the question of fairness for Chapter 18.

How the Untax Works

A carbon untax (or tax) is simple because it collects revenues from very few players. For example, an oil tax does not charge 200 million drivers every time they buy gas. And it does not tax tens of thousands of gas stations. It simply charges oil refineries for the amount of carbon in the oil they buy. Taxing oil refineries, natural gas producers, and coal mines would cover almost all carbon.

Refinery operators will, of course, complain about being taxed and forget to mention they are passing the tax on to gas stations. Gas station owners will complain and forget to mention they are passing the tax on to consumers. So when you hear their complaints, remember who really pays the carbon charge— it is you and I, the final consumers, and no one else.

When truckers buy gas, they will claim to be consumers because they burn the gas in their trucks. But, in fact, they will pass the cost on in their trucking rates. Anyone who can pass the cost on will pass it on, and if they pass it on they are not a final consumer. When you buy gas for your car, unless you can bill someone else for your gas costs, you are the final consumer. In essence, you pay the carbon tax.

I do not intend to discourage a carbon tax or untax by pointing this out; rather, I am encouraging self-defense. Even though businesses will pass the cost of the untax right through to us, they will demand a slice of our refund checks in addition. In fact, the cap-and-trade laws before Congress, which are basically disguised carbon taxes, include long lists of who gets how much of the tax revenue. And let me tell you, you are scheduled to get little to none. That's right. You pay the tax, and business gets the refund.

It's important to remember that even though the government collects the money from refineries and coal mines, you and other consumers ultimately pay the full charge. So the refund belongs to you—or at least it should. All 100 percent of it. I hope I am making myself clear on this, because when it comes to big bucks—and we are talking about hundreds of billions here—business is going to fight hard and fight dirty.

All right, let's look on the bright side. Say we win that fight and secure the refund for consumers. How does the refund work? It's simple. I suggest we do as Alaska does. Everyone who has been a legal resident for the past year gets a check in June. How big a check? Count the revenues for the last year and divide by the number of checks. Everyone gets the same amount.

Alaska spends less than 1 percent of the money it returns on mailing out the checks. The overhead should stay low because everyone will want to cooperate—if they don't, they don't get their checks. It's a lot easier to find people when you're handing out money than when you're collecting it.

How Big Is the Refund?

A standard guess at how high a carbon tax needs to be, at least for the next decade or so, is $30 per ton of carbon dioxide, though guesses vary widely. The United States emits about 6.5 billion tons of carbon dioxide per year (22 tons per person). At that rate, the untax would collect about $195 billion per year. The U.S. population is about 300 million, so that generates a refund of $650 per person, or $2,600 per year for a family of four.

An oil price of $100 per barrel is probably high enough on its own to encourage conservation, so the untax rate on oil might start out near zero (see Chapter 19). Coal and natural gas would still be taxed. This would reduce refunds to about $365 per person, or $1,460 for a family of four.

On average, everyone pays the same in higher prices as they get back in refunds, so this is not a get-rich-quick scheme. However, as I've mentioned, the very rich use more energy—heating their mansions and flying their private jets, for example—than do most of us. In fact, the rich use so much more carbon than average that they raise the average to a point at which 60 percent of the population uses less than the average. Everyone using less than average gets refunds greater than their additional costs of energy.

Energy Policy Number One: The Carbon Untax

The world is at risk of costly climate change, costly oil-price spikes, and more wars over oil. But contrary to what many believe, scientists do not yet know if a climate-change tipping point exists or where it is if there is one. Terrorist activity and wars are equally hard to predict. Action is clearly warranted, but we cannot pin down just how much to spend.

A simple realization provides the key to sensible action. After thirty-five years of complex and ineffective energy policies, the country was importing a greater percentage of oil, faced the highest oil prices ever in 2008, and was emitting more carbon than ever. It would be beneficial to put in place a solid, simple, efficient policy that could be dialed up or dialed down as needed. To achieve this, the policy should start gradually to overcome reasonable (and unreasonable) concerns about cost, but it should be set to ramp up unless it causes problems or we discover a magic energy technology.

The carbon untax is such a policy. It would be gentle and powerful at the same time. Most importantly, it would end thirty-five years of ineffective policies and prepare the country for the challenges ahead. Because the carbon charge part of a carbon tax is the same as the carbon charge part of an untax, we can turn to other experts for opinions on designing the carbon charge.

The most effective action would be a slowly increasing carbon
tax.

—Climate scientist James E. Hansen, 2006

Taxes on carbon are powerful tools for coordinating policies and
slowing climate change [and] are likely to be more effective and
more efficient [than] quantity oriented mechanisms like the Kyoto
Protocol. … Carbon prices would rise by between 2 and 4 percent
per year.

—Economist William Nordhaus, 2005

A carbon tax could be relaxed [or] increased. In either event, such
changes could be phased in over time, creating predictability and
allowing an ongoing reassessment.

—American Enterprise Institute, 2007

James E. Hansen, a NASA scientist, has long been the best-known and
most outspoken scientist warning of climate change. William Nordhaus, a Yale
economist, has been perhaps the leading energy economist for thirty years.
The conservative American Enterprise Institute has been skeptical of global
warming though concerned about energy-security issues.

Again, it is remarkable to find such a diverse group not only advocating the same policy, but describing its implementation in similar terms. Only
Hansen is advocating an untax, but the others recognize the political difficulties
of imposing the new tax they advocate.

A plausible path for the untax rate would be to start at, say, $4 per ton
of carbon dioxide in 2010—or as soon as possible, in any case—and increase
by $2 per year toward $40 in 2028. I would prefer a faster start, if it turns out
to be politically feasible.[1] We should commit to following the path we adopt
for, say, four years at time, and as the American Enterprise Institute report
suggests, changes should be phased in over time, not implemented suddenly.
A predictable approach will both save billions of dollars and accelerate the
impact of the policy by many years.

Here's how I recommend implementing a carbon untax:

- ► Start with a low carbon charge and increase it gradually.
- ► Apply the charge to all fossil fuels but collect it at the fewest possible
  upstream points.
- ► Mail checks to consumers in June that refund 100 percent of collected
  revenues on an equal-per-person basis.
- ► Reassess the carbon charge regularly but change it only gradually.

---

1. Notice that we say "carbon" tax, but the dollar values are actually applied to carbon
dioxide. A $12 tax on carbon dioxide is the same as a $44 tax on carbon.

Because of the slow start, those most concerned about the climate will undoubtedly worry that this is too little too late. But remember two points. First, climate-change advocates have been in a rush for almost two decades, without making any progress on emissions that can be measured in the atmosphere. A slow and steady start on a powerful policy is better than a continued deadlock or throwing money at wasteful policies like solar roofs and ethanol.

Second, don't forget lookahead. A predictable tax or untax rate that will only take effect in, say, ten years starts working as soon as it can be predicted. As an example of how this works, consider a new lending policy adopted by Bank of America. The *Wall Street Journal* discussed it in February 2008: "Bank of America says it has decided to start factoring a cost of carbon-dioxide emissions into its decisions about whether to underwrite debt for new coal-fired plants. Specifically, the bank says it anticipates a federal cap that would require a utility to pay between $20 and $40 for every ton of CO_2 its power plants emit."

Has a new green consciousness seeped into the Bank of America? No, it's still chasing the old-fashioned green. To make safe investments, the bank will assume a carbon permit cost or carbon tax of roughly $30 per ton, even though no such law has been passed. That's pretty amazing. The law has not even been drafted, and it's already working.

Bank of America is looking ahead at likely trends. If a nonexistent law can have such a strong financial effect in the present, so can the future tax rates of an actual law. Any scheduled increase in the carbon charge will have an impact long before it takes effect. A scheduled tax rate of $30 in 2025 affects coal plant investment decisions today.

The benefit of starting the tax slowly is that it is gentle in its effect on existing businesses, giving them time to adjust. This means less resistance from businesses and less need for handouts to get their buy-in. Nonetheless, if a quicker scaling-up of the untax rate gains enough popular support to pass, it will not do any significant harm to the economy and would benefit energy security and climate stability.

Why the Untax Works

As I just explained, consumers pay 100 percent of the untax and get it all back in refunds. At first, many people think this is nuts. But that's because they don't stop to think that, in the untax race, some consumers are winners and some are losers. Use less carbon, and you can be a winner, paying less than you get back in your refund check. Use more carbon than average, and you lose. That's why an untax works. Most people want to be winners.

It's the refunds that cause all the confusion. Sure, the carbon charge makes people want to use less carbon, but won't people spend all of their refund checks on paying the extra carbon charges? They could do that, but they will quickly learn that it's a waste of good money. Keep in mind, the refund does not change when you spend more or less on carbon. Suppose a family of four gets a $4,000 refund, no matter what. Suppose that with the new untax, it suddenly becomes possible to save $800 a year—all costs included—by installing more home insulation. You could say, "Why bother? I've got my refund check to spend; I don't need to save $800. I can just pay the higher energy bill." A few may say that the first year, but then it will sink in: Why send $800 of my refund check to my utility company?

In the end, the tables will turn, and most people will decide it's nuts to treat their refund checks like burnt offerings to their local utility companies or gas stations, just because the money came from charges on carbon. Who cares where the money came from? No need to spend it all on carbon taxes. Consumers will find ways to cut back on fossil fuel and spend the checks on their own needs and desires.

If the Refund is 100 Percent, Is the Untax Free?

The untax works in spite of returning every penny collected. Direct costs—the total paid to the government less the refunds from the government—sum to zero. But does this mean the untax is free on average? No. If the untax works and gets people to do things that reduce emissions, the untax causes indirect costs. Indirect costs—which I also call hidden costs because people often either don't notice them or ignore them—are what people pay to get the job done. Hidden costs don't show up in untax accounting, but they are the real costs of carbon policy.

Buying a hybrid car because of an untax provides one example of hidden costs. Suppose buying the hybrid would cost you $3,000 extra but would save you $2,800 in gas cost over the life of the car. The net real cost to you of using the hybrid is $200, so you don't buy it.

Now suppose we impose an untax, which makes gas more expensive, so buying a hybrid saves $3,200 on gas. Now it saves us money to buy a hybrid. But, not counting climate or security benefits, there is still a net social cost to buying the hybrid. It still saves only $2,800 worth of gas, and we only bought it because it also saves $400 in untax payments. The untax has tricked us and rewarded us into spending $200 more on a hybrid than we save on gas (not counting the untax savings). This is the real, but hidden, cost of the untax. We don't see it because we're getting rewarded by the untax refunds.

Spending more than the true savings would make no sense, except that there's an extra benefit to using less oil that we're not counting—climate stability and energy security. That's what we get for paying the hidden cost. Carpooling provides another example of a hidden cost—the cost of inconvenience. No dollars change hands over inconvenience. But it's still a real cost.

The true cost picture shows that the hidden costs are real, and the obvious direct costs, which everyone discusses, net out for society as a whole to nothing at all. But the direct costs—the carbon charges—cause people to save carbon. Saving carbon often does cost money, and these hidden costs are hard to keep track of and are usually overlooked. But in one case, when they are zero, they are easy to count. If no one does anything to save carbon, there are no hidden costs and no net cost to society. That's worth remembering.

▶ If the untax fails to cause any conservation, it's completely free when averaged over all consumers.

The more good an untax does—the more it reduces fossil-fuel use—the greater the hidden costs. But there's a limit. In the hybrid example, people saved $400 on gas costs because of the untax. That tells us something about the hidden cost. It cannot be more than $400 per person, because people are smart. If the hidden cost of switching to a hybrid was $2,000, they would not do it to save $400. This puts a strict limit on the hidden costs.

▶ If the untax works, the maximum possible hidden cost is the amount of carbon charge (tax or untax) avoided.

This just tells us the maximum possible cost in the most extreme case. Typically, the hidden costs are much less. People conserve in the least expensive, least inconvenient ways first. In fact, the first bit of conservation is typically almost free. Economics shows that hidden costs are typically only half the maximum possible value.

▶ The typical hidden cost of an untax is half the amount of carbon charge avoided.

Using these standard results, I have calculated the approximate hidden costs of an untax with various carbon charges and various levels of effectiveness (see Table 1).*

Table 1. *Average Total (Hidden) Cost per Person per Year*

Charge per ton CO_2	Percent CO_2 Abatement Caused by Untax			
	10%	20%	40%	80%
$4	$4	$9	$18	$35
$10	$11	$22	$44	$88
$30	$33	$66	$132	$264
$60	$66	$132	$264	$528

Based on emissions of 22 tons of CO_2 per person per year before the untax.

Table 1 contains good news and is, in fact, much of the reason that economists favor a carbon tax. It says that imposing a $30 carbon tax, which has a direct cost of $528 per person per year, would only have a real cost that averaged $66 per person per year if it cut carbon emissions 20 percent. The direct costs, which receive all the publicity, come to $528 per person, but net to zero counting refunds. The real net costs are eight times less. That's why checking the economics is so important.

The table is based on a very simple approximation. So I checked it against the results of the complex economic model of cap-and-trade costs used by researchers at the Massachusetts Institute of Technology. Except for the first few years of their model runs, the results were quite similar. In all cases my simple approximation indicated higher costs than the more rigorous MIT model.

Of course, the untax cannot be designed to save 80 percent with a carbon charge of only $4 per ton. Only the carbon charge can be set, and then an abatement level will occur on its own. A 20 percent abatement in response to a carbon charge of $30 per ton is quite plausible, but only time will tell. The $66 real cost of such a policy is roughly the cost of one tank of gas per year. This is why a good energy policy just cannot wreck the economy.

Even in a relatively extreme case, which we would not encounter for decades, the $520 cost per person per year is barely over 1 percent of gross domestic product (GDP). Of course, decades from now, GDP will be higher, and energy use per GDP will have fallen considerably.

Impact on the Poor. A cost of $66 per year is more difficult for the poor. But this is the average cost, and a person with a very low income is unlikely to be an average user of fossil fuel. The poor don't own private planes, don't fly very much, and don't heat big homes, swimming pools, or hot tubs. If they used just 20 percent less energy than the average user, and did not bother to conserve at all, they would come out ahead on refunds by $104 per person per year. They would come out ahead by more if they took any energy-saving action that they found worth the money.

Impact on Oil Prices. One more thing to remember is that all energy policies that cause a reduction in oil use will lower the world price of oil. The effect on oil prices will be doubled or tripled if such policies become the basis of the next international climate policy. That would save the United States a lot of money. In fact, it could save enough to cover the entire real cost of the untax by, in effect, charging it to OPEC.

~

The untax is the silver bullet of energy policies. It's simple, fair, and efficient. Most of the best policy experts advocate it, and most politicians fight it. The trouble is the T word. But, as the smartest conservatives are saying, we need to be free to discuss taxes. Demonizing the word *tax* wins votes but forces the country into more costly policies. Ironically, the likely alternative—cap and trade—is simply a cleverly disguised carbon tax ultimately paid by consumers but largely refunded to polluters.

The untax is administratively simple and cheap because it collects the carbon charge at the fewest possible points, and all refunds are equal. It is fair because it rewards those who do less harm than average—about 60 percent of all families—and places a net charge on those who do more than their share of harm. Yet it is not dictatorial. Everyone is free to burn as much carbon as they can afford. But almost everyone will choose to burn less.

The untax is powerful and efficient because it is a true market mechanism. It simply raises the price of carbon to the level it would be if the market worked perfectly and included the costs of all side effects. It reaches into every corner of the economy that uses carbon and provides an incentive to use less. It is powerful for exactly the same reason that OPEC's energy experiment from 1973 to 1985 permanently transformed the world's use of fossil fuel and saved a hundred times more carbon than any other policy before or since.

chapter 17

Untaxing Questions

It seems to me a bit like buying indulgences from the ancient church. ... I can waste all the energy I want and then justify it by writing a check.*

—Former Arkansas Governor Mike Huckabee, 2007

THE CLIMATE IS CHANGING. The terrorists are coming. We've got to do something now. Grow more corn. Make hydrogen. Build nuclear reactors. Build solar roofs. Cap greenhouse gasses. Invent fusion reactors, zero-emission vehicles, nanotech this, and biotech that.

These ideas all sound so concrete and effective. But sound is about all we get. Ethanol makes things worse, the hydrogen bubble has burst, and zero-emission vehicles zeroed out. Still, there will always be new energy fads.

Carbon taxes and untaxes, on the other hand, are not fads. But it's hard to put your finger on just what they do. They quash the fads and accelerate ordinary, but effective, conservation and give wings to real breakthroughs. But I can't predict the breakthroughs, so it's hard to make an untax seem sexy. Still, perhaps I can at least rebut a few of the baseless criticisms that will surely hinder its acceptance

Indulgences from the Ancient Church?

Both carbon emission permits and a carbon untax let polluters buy their way out of the energy policy. If you have the money, you can emit as much as you want—or even more just to be spiteful. This strikes many people as immoral, so they dismiss market-based policies. As Huckabee puts it in this chapter's opening quote, "I can waste all the energy I want and then justify it by writing a check."

Although as an economist I should probably not admit this, I feel much the same way. I dislike seeing the rich abuse the environment for selfish reasons. In spite of this, I favor policies that let them do just that. My motive is practical. I have taken a close look at every way I can think of—more ways than I discuss here—to curb rich polluters, treat the poor fairly, and still make large cuts in oil use and carbon emissions.

I see no way to do all three. This requires a choice, and my choice is to curb carbon emissions and treat the poor fairly. The rich are beyond our control, so I say we should at least sell them indulgences. But let's not give the money to the ancient church—or to the modern government either.

But why can't we force the rich to do their part? If we imposed a 30 percent cut in carbon use on everyone—no exceptions—the rich could not wriggle out of that. It does seem unfair to the poor, who are already getting by with very little. But the real problem is that it can't be done. How could we count up everyone's carbon every year? Heating, driving, flying, boating, lighting—how could we count all that for every person? It's just impossible. If you can't count it, you can't cut it 30 percent. The same problem applies if you require everyone to reduce their carbon use by the same number of tons. Plus, it would devastate the poor and not make much difference at all to the rich.

Since we can't keep track of everyone's carbon use, perhaps we should keep track of everything else. We could require that all cars get at least 30 miles per gallon. We could ban through-the-door ice makers on refrigerators, because they waste a lot of energy. We could restrict carbon use for heating and cooling to 1 ton of carbon per year per house. Or, if we don't like this one-size-fits-all approach, we could set a different limit for each size of house in each part of the country. But how many miles of plane travel and driving should we allow? Obviously, this approach is a nightmare of regulation.

It is possible, though not a good idea, to use command-and-control regulation on large industries, but when it comes to individuals it really does not make sense. The problem is that energy use reaches into every corner of our lives. Controlling the rich would require the government to check every corner. No one thinks that's right, and fortunately, it's completely unnecessary. We can actually do something that's fair to both the rich and the poor—and that's the untax. It lets the rich write checks, and when the refunds are given out equally the poor get back more than they pay. I explain, in the next chapter, why this is exactly fair.

Do Consumers Care about Price?

A related objection to the untax is that it won't be only the rich who ignore it; everyone else will as well. Everyone is so addicted to fossil fuel, the thinking goes, that they will pay whatever it takes to get their fix. This is the pop-psychology approach to economics. Economists go a bit overboard assuming people are rational, while pop psychology sees people as irrational—but predictable. I'm as skeptical of predictable irrationality as I am of rationality. It's best to take an experimental approach to human behavior. Fortunately, the experiment has been done.

Looking back at the OPEC crisis, we find that OPEC had its effect on the world entirely by means of price. (OPEC didn't cause the lines at the gas stations, by the way. Misguided regulation in the United States did that.) Yes, OPEC pumped less oil, but the world oil market did what markets do and made sure that anyone could buy as much oil as they wanted—provided they paid the price. Think about this for a minute. OPEC supplied less oil, but that did not stop anyone from buying more oil. In every case, it was the price that stopped people from buying more. Prices tripled then doubled on top of that. Price changed behavior, and the change was enormous.

Look back at Figure 3 in Chapter 8. It shows total U.S. energy use before and after the OPEC crisis. The figure is not a product of green conservationists, but of Dick Cheney's National Energy Policy Development Group—an organization dominated by energy supply companies. If the OPEC crisis had not occurred, the United States would have used something like 165 quads of energy in 2000, according to Cheney's group. Instead, the United States used 100 quads that year—a savings of 65 quads. (A quad—for quadrillion British thermal units—is a whole lot of energy.) The data includes energy from all sources—fossil, nuclear, and alternative. So this graph shows just one thing—the effect of high prices on total energy use. The total energy saved is equivalent to almost two decades of oil use at the 2007 rate. In 2007, the United States used only 40 quads of oil.

Some will argue that this enormous impact is due to fuel-efficiency standards and other government programs. But consider two points. These programs would never have happened without the OPEC price hikes, and the government programs do not account for the bulk of the effect. In fact, when the Department of Energy checked energy impacts in 1980, it found that government programs had had almost no net impact on energy conservation but that price had been effective in encouraging energy savings (see "Energy Policy: Mostly Sound and Fury" in Chapter 7). The high prices imposed by OPEC saved vastly more energy than any other policy before or since.

But would that work again with an untax? The untax could be set to mimic OPEC's price increase. The difference would be the refund checks. When

OPEC raises the price and sends no refund checks, we have two reasons to use less oil. The price of gas is higher, and we end up poorer as a nation because of all the money we give to OPEC. If we do it ourselves, with an untax, the high price has the same effect, but since the untax revenues are all refunded, the country doesn't get poorer.

Which effect is more powerful, the high prices or becoming poorer? When OPEC raises the price of oil from $25 to $125, that's a 400 percent increase. But that makes us only about 5 percent poorer as a nation. Not surprisingly, economists find that the price effect is much larger than the income effect. Since the main effect is from price, the untax works almost as well as having OPEC raise the price. Of course if we wanted the untax to work just as well as OPEC, we could make it a tax and then give all the revenue to some other country. That would make us poorer, and we would buy less fossil fuel. Not a great idea. In fact making ourselves poor to conserve on energy is probably the worst possible energy policy—and that's the only advantage that OPEC's tax has over an untax.*

Because the United States is wealthier now, high carbon prices will probably have less effect than they did thirty years ago. But being wealthier is not a bad thing. On balance, it helps more than it hurts and will make the transition to alternative fuels easier than it would have been in the past.

What's the Psychology of the Untax?

The objection that everyone will ignore the untax because they are addicted to fossil fuel—that the untax is too small to matter—is based on a view of people that lumps them into a few types, sometimes even just two types. For example, some people like SUVs and some like small cars. The SUV owners won't switch to small cars, and the small-car owners already have small cars, so a carbon tax won't do much good—or so the thinking goes.

People—and even cars—are far more complicated. There are a hundred types of cars and a hundred million types of people. Think about an election with two candidates. The polls tell us that 40 percent of voters favor Sue Spender and 50 percent favor Tom Taxer, with 10 percent undecided. Will one bad headline for Taxer have no effect, because people are either for him or against him? That's probably true for 70 percent of the voters. Their minds are made up, and it would take a lot to change them.

But election strategies are all about shifting the fence-sitters, and, invariably, about 10 percent of voters are on the fence—undecided—or extremely close to the fence. For them, little things can make the difference. And notice that once the bad headline shifts the fence-sitters to one side, a new group of fence-sitters climbs on. A new poll might say that 45 percent favor Spender and 45 percent favor Taxer, and there are still 10 percent on the fence.

The same is true for consumers. On every energy decision—whether to buy a smaller car or a better furnace—most people are firmly in one camp or the other. They already have a smaller car, or they definitely don't want one. But it's wrong to think of people as coming in just two types. Even if a lot are in one camp and a lot in the other, you always find a good number sitting on the fence. These are the people who respond to the first small change in fuel price. And once they respond and move off the fence to the low-energy camp, a group of people who use more energy move onto the fence, ready for the next price increase to shift them.

Because we face thousands of energy choices, most of us end up on the fence for at least some choices. And if I'm not on the fence now, I may well be in five years, when I need a new car anyway.

Should I drive to the store or walk? It depends on the weather, which is sometimes borderline, so I'm on the fence. Should I turn off more lights or buy more compact fluorescents or check the air pressure in my tires? With higher energy prices, I will think about all this a bit more and make some of these choices differently. Human psychology is not often black or white; in fact, it's tremendously variable. Changing the price of carbon shifts the weights on every decision, and choices that are at a tipping point will tip away from carbon.

The power of the untax is that it shifts the weight on so many billions of choices that it gets maximum bang for the buck. The beauty of the untax is that it shifts only fence-sitters or those close by. These are the people who are bothered least by making a change. So a carbon untax (or a carbon tax) makes all the changes that bother people the least. Those who really don't want to change don't have to. If they are big energy users, the carbon untax charges them, and if they use only a little, the untax rewards them.

Is the Untax Good for Alternative Energy?

As the OPEC crisis demonstrated, the main response to higher oil prices is conservation. Because of the crisis, non-OPEC supply increased a bit, but demand dropped by about ten times as much. The point to understand about the untax, or any type of energy price increase, is that it is not aimed solely at conservation. It induces more conservation only when conservation is easier and cheaper than the alternatives.

The untax targets all nonfossil fuels just as strongly as it targets conservation. In fact, it also targets all innovations and inventions for new types of conservation and new sources of alternative energy. No regulatory policy could do that. So even if you don't believe conservation is possible, the untax still does the job. This is important, because in the long run, the world needs new technology. The untax will encourage all possible new technologies equally and make sure we get the cheapest one.

~

An untax rewards not only conservation, but also alternative energy production and all types of innovation. In fact, it creates a level playing field for all alternatives, present and future, to fossil fuel. But only the cheapest approaches will win out.

A carbon tax derives its power from its breadth. It puts a little extra pressure to use less on every fossil-energy decision, something no other policy can accomplish. It doesn't alter most energy choices. It just changes decisions near a tipping point that is sensitive to a change in cost. That's how a small push has a large effect.

Some say that taxes and untaxes are so weak that people will ignore them. Considering that "no new taxes" is the most potent of all political slogans, it seems odd to think that people would ignore taxes, and in fact history shows they do not. OPEC's "tax" caused great outcry, and, unsurprisingly, the historical record shows a massive and permanent change in the world's use of fossil fuel. Never underestimate the power of tax avoidance.

chapter 18

Why Untaxing Is Fair

*The guys with money will still be able to afford as much gas as they want.
Only the little guys will suffer.*

—Rita Gibson, Boston delicatessen owner, 1977,
quoted in *Time* magazine

"SLAP A 5¢-PER-GAL. TAX ON GASOLINE each year if conservation goals are not met." That's how *Time* magazine described President Jimmy Carter's proposed gas tax shortly after he took office and declared the energy crisis to be the "moral equivalent of war." But people had adjusted to OPEC's tripled price and were getting complacent. No one foresaw that the Iranian revolution would soon trigger a doubling of the already high oil price.

Intense lobbying by the oil and gas industries derailed Carter's proposals, but America's sense of fairness also played a role. Carter saw that, higher though they were, oil prices were not yet high enough. And he proposed several corrections, one of which was the five-cent gas tax. That's similar to the carbon tax I've been discussing. Taxes are never popular, but the gas tax struck people as particularly unfair, and they were right.

According to the Congressional Budget Office, a carbon tax would cost the poorest one-fifth of families twice as much in terms of percentage as those in the upper fifth. The low-income group emits only a third as much carbon as the high-income group but suffers more under a carbon tax. Rita Gibson was right: "Only the little guys will suffer."

Many economists recognize the fairness issue and attempt to solve it with some form of tax relief. Harvard economist N. Gregory Mankiw, for example, advocates a "rebate of the federal payroll tax on the first $3,660 of earnings for each worker." Such a tax rebate would distribute the carbon tax revenues in a way similar to the untax refund, so in spirit Mankiw is close to my position. But as I will show, his carbon tax with payroll tax reduction is not quite as fair as the untax. And as the headline of an op-ed he wrote for the *New York Times* proclaims, it's "a new tax"—a huge new tax that will never fly.

Mankiw's op-ed captures the economist's dilemma perfectly. It's about the extreme difficulty of passing a carbon tax, simply because it's a tax. But the headline emphasizes only this problematic quality. Why is Mankiw beating his head against this wall? Why not suggest refunding the tax revenues, turning his new tax into an untax? Is the untax so novel an idea? Hardly. Economists habitually model a carbon tax as an untax. It's an old and venerable idea. So why avoid it? Because economists think they have an even better idea.

Most economists believe that using the carbon-tax revenues in place of regular tax revenues is better, because it is the most efficient approach. They say this approach provides a double dividend: we use less carbon, and taxes are more efficient So politics be damned. These economists want to recommend the best approach—even though they know it is political suicide. I admire this insistence on doing things efficiently, and for twenty-five years I bought the standard analysis that using the carbon-tax revenues in place of other tax revenues is a great idea. But this chapter shows it's not, and that's a great relief. There's no need to keep banging our heads on the no-new-taxes wall.

But could most economists really have missed this point for so many years? Yes, and for a reason. According to economics, we should judge a carbon tax or untax on two counts: efficiency and fairness. Efficiency just means cost-effectiveness. Fairness concerns taking money from one group and giving it to another. Unfortunately, fairness is usually difficult to assess, so economists usually ignore that issue and focus instead on efficiency. Economists have done just that with the carbon tax, proving that Mankiw's approach is a bit more efficient than an untax. Efficiency is the sole reason Mankiw and other economists bang their heads on the no-new-taxes wall.

But a complete comparison between a carbon tax and a carbon untax requires considering fairness as well as efficiency. I have never seen anyone attempt this, but I will in this chapter. By a stroke of good luck, it turns out to be possible. I say good luck because I know of only one other policy that economists agree is wrong because it is unfair, even though it improves efficiency. Let's call it policy X. Surprisingly, policy X is exactly the difference between a carbon tax and a carbon untax.

In a nutshell, this chapter shows that an untax is completely fair and that a carbon tax is just an untax plus policy X. Since economists agree that policy

X is wrong, they should agree that using the completely fair untax plus policy X is worse than using just the untax.

The Gold Standard of Fairness

The problem of global pollution resembles an old economics puzzle called the "tragedy of the commons." Of several possible solutions to this puzzle, one stands out as the most obviously fair. The tragedy of the commons refers to the story, with some basis in reality, of an English town's common pasture for grazing animals; let's call them sheep. Everyone can graze as many sheep as they like on the commons at no cost. The tragedy is that everyone takes advantage of this free resource, overusing it. Overgrazing kills the grass, and the commons becomes nearly worthless.

Global warming parallels this story in several ways. People can dispose of their carbon dioxide in the atmosphere for free, but the carbon dioxide reduces the value of this common resource. Economics suggests a solution to this problem—a solution that is widely agreed to be fair, although impractical on a global level. I will show that the untax is a practical way of getting this same fair result. But first, let's look at the standard fair solution.

To avoid the tragedy fairly, the town determines how many sheep the commons can sustainably support and divides this number by the number of townspeople. Say it comes to two sheep per person. The town grants each person the legal right to graze two sheep. That's fair, and it prevents overgrazing. To an economist, this is also an efficient solution because it maximizes the value of the commons. Any more sheep, and they would damage the commons. Any fewer, and the town would not fully utilize its commons.

If, however, the blacksmith does not want to graze sheep, a fair solution allows him to give away his rights, trade them, or sell them. After all, he should have the right to do what he wants with his rights. That's why we call them rights.

In a large town, a market for rights to the commons develops. It likely just consists of a bulletin board with notices. But soon a typical price develops for, say, the right to graze one sheep for a month. That becomes the market price of sheep permits. In this way, the blacksmith can sell his right at fair market value, and neither buyer nor seller takes advantage of the other.

Notice that we have just reinvented, probably for the millionth time, the system we call cap and trade. The town caps the number of sheep at the sustainable limit of the commons and gives all the townspeople permits, which they are allowed to trade.

This system provides a fair and efficient solution to the tragedy of the commons. Giving out rights equally makes the system fair, because no one has any special claim to extra rights. Giving out a sustainable number of rights and allowing trade makes the system efficient.

Conceptually, this system provides a fair way of solving the problem of climate change. Unfortunately, while fair in principle, giving people such rights and enforcing them on a global scale would be impossible.

Fairness: Twin Gold Standards

A fair solution to the global commons problem requires a special cap-and-trade system. Typically, carbon cap-and-trade systems work by distributing rights not on an equal-per-person basis, but in proportion to how much damage each person was causing on some past date. The big polluters get the rights, and as Gibson predicted, the little guys suffer.

That's not a fair cap-and-trade system. The only fair system for handing out rights is to give them out on an equal-per-person basis. I will call this particular cap-and-trade system *equitable* cap and trade.

This system should sound familiar. The untax gives refunds on an equal-per-person basis. The untax and equitable cap and trade are twins—provided they are adjusted to give the same carbon price. These prices are the same if permit prices under the cap are the same as the tax rate under the untax. Emitting carbon has the same cost in the two systems, so people reduce emissions by the same amount. The free permits, given out equitably, benefit low carbon consumers exactly as do untax refunds. I explain the reasons for this in "Why the Untax and Equitable Cap-and-Trade Are Twins" for those who wish to delve deeper into the economics.

The two systems differ in only one way: The market sets the price of permits, so their cost fluctuates unpredictably. Still, on average, costs, revenues, and emissions all come out the same, so the two systems must be equally fair. This makes the untax a twin gold standard of fairness. It treats people as if they had equal rights to the climate, but without keeping track of 6 billion individual climate rights.

Enter the Economists

The untax is as fair a system as anyone can devise without getting into person-by-person calculations. Because such calculations add enormous complexity and are difficult to make fair, they should remain outside the untax system, even in the few cases in which they are practical.

In spite of the fact that the untax is the fairest system for correcting the underpricing of carbon, many economists recommend against the untax. Princeton economist, *New York Times* columnist, and Nobel Laureate Paul Krugman says that "any new tax on carbon could and should be offset by

Why the Untax and Equitable Cap-and-Trade Are Twins

It's hard to imagine individuals owning and trading carbon permits, so this example uses the analogy of a town commons with grazing sheep. To understand why an equitable cap-and-trade program and the untax are twins, consider two identical towns, one with an untax and one with equitable cap and trade. The permit price equals the untax rate; suppose they are both $10 per sheep per month. (Permit prices are not that stable, but let's keep the example simple.)

Suppose each town has 100 people, and Cap Town caps the number of grazing sheep at 200. In Untax Town, with no cap, let us guess that citizens are grazing 220 sheep and paying taxes on them as well. We will soon check this guess. The per-person refund comes to $10 times 220 sheep divided by 100 people, or $22 per person per month.

	Mary Untax			Jane Capper	
	Tax	Refund	Untax Cost	Free Permits	Permit Cost
1 sheep	$10	$22	− $12	2	− $10
2 sheep	$20	$22	− $2	2	$0
3 sheep	$30	$22	+ $8	2	+ $10

The table shows the choice faced by a pair of identical twins, one in each town. If Mary Untax grazes one sheep, she pays $10 in untax but receives the $22 refund, for a gain of $12. (The table shows a gain as a negative cost.) Jane Capper receives two free permits. If she grazes only one sheep, she sells one permit, for a gain of $10. Notice that, in both towns, it costs $10 more for each additional sheep grazed.

In either town, if grazing another sheep makes more than $10, it's a good idea. Otherwise it's not. So Mary and Jane will decide to graze the same number of sheep—and so will every other pair of twins in the two towns.

Originally, we were not sure how many sheep people would graze in Untax Town, but now we know. At the same cost of $10 per sheep, the two towns graze the same number of sheep. Since 200 sheep graze in Cap Town, 200 must also graze in Untax Town—not 220 as we first guessed. So the refund per person turns out to be only $20. Recalculating the table shows that the two towns match perfectly.

	Mary Untax			Jane Capper	
	Tax	Refund	Untax Cost	Free Permits	Permit Cost
1 sheep	$10	$20	− $10	2	− $10
2 sheep	$20	$20	$0	2	$0
3 sheep	$30	$20	+ $10	2	+ $10

tax cuts elsewhere." Mankiw would use the carbon tax to pay off some of the federal payroll tax.

Economists reason that taxes cause us to do less of what is taxed. Income taxes cause some to work a little less; taxes on capital cause people to invest a little less. Working and investing are beneficial and increase the gross domestic product (GDP), so taxes reduce the GDP. Even a carbon tax reduces GDP a bit, but economic calculations indicate that taxing carbon reduces GDP less than taxing labor or capital.

This means that replacing part of the tax on labor or capital with revenues from the carbon tax would increase GDP. That's a good thing, but how big is the effect? Dale W. Jorgenson, the statistical economist we met in Chapter 2, has answered this question. He estimates that swapping a carbon tax for taxes on labor would increase GDP by 1 percent and that swapping a carbon tax for taxes on capital would increase it almost 3 percent. These results apply to a carbon tax that cuts emissions by 30 percent.

So the effect of using carbon-tax revenues to pay off other taxes is beneficial, but not too impressive. Consider the 3 percent gain from reducing the tax on capital. GDP grows by about 3 percent every year, so after twenty years with compound growth we might be 83 percent richer instead of 80 percent, if economists got to swap the carbon-tax for a tax on capital.

So the economists have a point. If they don't let us have the refund and they instead use the $300 billion or so per year of revenues to reduce taxes on capital, we will end up a bit richer on average. That's why economists want to make the carbon tax a new tax—to replace an old tax.

Reenter Fairness

But *richer on average* doesn't say what happens to you or me individually. Perhaps you will lose 10 percent and I will gain 16 percent, and so we will be better off by 3 percent on average. A lot of good that does you.

Economists know that they should take fairness into account, and, strictly speaking, if something is better only on average, economists should not say the situation is better. But they get frustrated when they can see that a policy improves GDP, but they don't know how fair the policy will be. So they figure, let's bet on the part we understand and cross our fingers that the other part—the fairness part that we don't understand—doesn't cause too much trouble. That's not a bad rule of thumb.

But in the case of an untax, it's possible to evaluate fairness conclusively, although economists have overlooked this fact. Let's take a look. In particular, let's look at the idea of using carbon-tax revenues to reduce some other tax, which I'll call tax T. Is reducing tax T a good idea?

To answer this question, we must consider the deplorable policy X that I mentioned at the beginning of this chapter. Policy X is known to economists as a *poll tax,* which is an old English term, or as a *capitation tax,* which means a tax on heads. That's a tax that charges everyone the same amount, no matter what. For example, the poorest person is taxed $1,000, and the richest person is taxed $1,000. No one I know approves of such taxes anymore, and I have never heard of an economist recommending that a capitation tax be used to raise revenues to reduce taxes on labor, capital, or anything else.

But economists agree that replacing other taxes with a capitation tax would increase economic efficiency and increase the GDP.[1] So rejecting the capitation tax means they believe that the unfairness of such a tax overwhelms its benefits. That's a judgment I think we all share. Replacing current tax revenues with a capitation tax is simply too unfair and should be rejected.

Now consider three policies, each of which collects and distributes an average of $100 per person:

#1. A carbon tax used to reduce tax T.
#2. An untax with an equal-per-person refund.
#3. A capitation tax used to reduce tax T.

Could number one, the economists' new carbon tax, be better than number two, the untax? To answer this question, consider an easier question, which turns out to be the same question in disguise. If we had policy number two, a $100-per-person untax, would it be a good idea to add to it policy number three, a $100 capitation tax used to reduce tax T?

Is policy #2 + policy #3 a good idea?

No, because we have already seen that using a capitation tax is so unfair that everyone rejects it even though it increases efficiency. There is no reason to change our minds and start liking capitation taxes just because we have implemented an untax—the fairest form of carbon tax.

But an untax plus a capitation tax—number two plus number three—is exactly the same as number one, a carbon tax used to reduce tax T.

Policy #2 + policy #3 = policy #1.

Here's why: Start with policy #2, the $100 untax. It's just a carbon tax with an equal-per-person refund of $100. Now add policy #3, a $100 capitation tax. That takes away everyone's $100 refund. So we are back to a regular carbon tax with no refund. That's policy #1.

---

1. The economic argument is that normal taxes discourage what is taxed, which means the taxes can end up discouraging something good. But a poll tax discourages nothing and so causes no good thing to be avoided. One cannot reasonably avoid having a head.

Because policies #2 and #3 together are a bad idea, and together they are the same as policy #1, then policy #1 must be a bad idea. And that's my point. The economists propose policy #1, a carbon tax used to reduce another tax. Compared with an untax, it's a bad idea.

The only way out of this logic would be proof that a capitation tax would cancel out some unfairness in the untax. But since the untax is one of the twin gold standards for carbon fairness, that way out doesn't make sense.

This conclusion is important because it removes a stumbling block that causes economists to advocate a new tax. They can control carbon emissions just as efficiently with an untax as with a carbon tax. And by turning the carbon tax into a carbon untax, they avoid the political pitfall of the T word. That should be comforting as well as familiar, because the untax has been showing up in economic models for years.

From Theory to Dollars

In 2000, the Congressional Budget Office (CBO) ran some numbers on the fairness issue. It considered a cap-and-trade system in which the government auctions off all the permits and uses the revenue to provide "each household with an identical lump sum." This is exactly the equitable cap-and-trade approach just discussed, except that it works per family instead of per person. But since equitable cap and trade is the twin of the untax, the CBO report can also serve as an analysis of the untax.

The CBO also considered a cap-and-trade policy in which all the permit revenues go to reduce corporate income taxes. This is just like a carbon tax whose revenues are used to reduce the corporate income tax. Table 1 shows the CBO's results.

Table 1. *The Congressional Budget Office Compared Two Ways of Using Permit Revenues or Carbon-Tax Revenues**

	Change in Real Annual Income	
Use of Carbon Revenues	Lowest 20%	Highest 20%
Decrease in Corporate Taxes (Efficient but Unfair)	−$510	+ $1,510
Equal-per-Person Refund (The Untax Approach)	+ $310	−$940

The table shows that when carbon taxes or permit revenues are used to reduce corporate taxes, the poorest 20 percent of households experience a net cost of $510, while the wealthiest 20 percent gain $1,510—in spite of producing more carbon emissions. With an equitable cap-and-trade program or an untax, *poor* families gain $310, and wealthy families experience a net cost of $940 per year because of their extra carbon emissions. These estimates apply to a policy that is intended to reduce emissions by 15 percent.

When the government reduces corporate income taxes with carbon-tax revenues, the poor, who emit less, get poorer, and the rich, who emit more, get richer. Gibson is right again.

Mike Huckabee complains, "I can waste all the energy I want and then justify it by writing a check." An untax allows such waste and check writing, but the checks written by the rich provide the poor with a small net gain to compensate them for climate rights they are not using and that, in effect, the rich are using. The cost of this compensation is enough that it will change the behavior of the rich without the government having to interfere with the details of their lives. Of course, everyone has the same incentive to emit less; no one is singled out.

Some of the rich will ignore the cost, and some will reduce their emissions significantly. All are free to choose their own strategy, but those using more than their share must compensate those using less than their share. If I were poor, I would rather the government charge the rich and send me some of the proceeds than heavy-handedly force the rich to cut back and give me nothing.

~

The untax, a carbon tax combined with an equal-per-person refund, has the same economic effect as giving everyone an equal right to emit carbon and allowing them to use or sell their rights. Although it may sound antisocial to "privatize the climate," the current system already privatizes the climate by allowing everyone to claim any amount of the atmosphere for their own private use without compensating anyone.

Redirecting untax refunds to reduce other taxes would increase economic efficiency a little. Reducing corporate taxes increases efficiency the most but is the most unfair. Any use of untax refunds to replace tax revenues turns the untax into a tax and is as unfair as implementing a capitation tax to reduce other taxes—a policy that almost everyone has rejected consistently for over a century.

The untax can protect the atmosphere to any desired degree simply by setting the appropriate tax rate. It allows us to choose how much to emit, but those who choose not to do their part must fairly compensate those who do more than their share.

chapter 19

Taxing Oil—Double or Nothing

Bush is dead wrong. … Vice President Bush was resolved on arriving in Saudi Arabia to plead with the sheiks to restrict the production of oil. … Mr. Bush would do better to announce to Sheik Yamani that … any oil coming this way … is going to cost X plus $10 per barrel.

—William F. Buckley, *Atlanta Journal*, 1986

WHEN GEORGE BUSH SENIOR, then Ronald Reagan's vice president, decided to help his friends in the oil business by nudging the price of oil back up, he knew what was needed. So off he flew to Saudi Arabia.* It was the Saudis who had, as William F. Buckley explained in 1986, "cost us something on the order of $400 billion or $500 billion." And it was the Saudis who had burst the price bubble at the end of 1985—not that they could have held out much longer, but they picked the time and opened the spigot.

Both Bush senior and Buckley understood that the Saudis, not American oil producers, controlled the price of oil and gasoline. However, in the short run, both Bush and Buckley lost. Bush urged the Saudis to restrict output and raise the price of oil, but the Saudis refused. Buckley recommended that the Reagan administration tax foreign oil to hold down the world oil price, and Reagan refused. Still, American oil interests won out in the long run. The absence of an effective energy policy restored OPEC's power, and beginning in the early 2000s prices returned to a level oil companies prefer.

For thirty-five years, grassroots American politics has gotten the whole picture pretty much backward, which is one reason we have made little progress in saving energy—except for the changes OPEC forced on us. Conventional wisdom holds that we need to fix "market manipulation" by domestic oil companies, tax their excess profits, and lower gasoline prices. I picked up today's newspaper, and every one of those issues was in it, but not OPEC. The same would have been true on a thousand days in the last thirty-five years.

But it's what's not in the news—OPEC and the world market—that matters most. Lower gas prices sound appealing, but if an addict is having trouble paying the high cost of drugs, should we make the drugs cheaper? An addict would think so. The only way to reduce oil addiction is to use less oil—pretty simple to understand, unless you're addicted.

Now, oil use is best reduced by high prices, and that's what confuses people. High prices reduce oil use, which causes low prices. So to get lower prices we need higher prices. No wonder people don't trust economics. But there's a method to this madness, and this chapter explains how to make the high-price method work with minimal pain.

I first resolve this seeming paradox by explaining that there are two different prices, the world market price and the domestic price. We raise the domestic price to lower the world price. Lowering the world price means Big Oil and OPEC get less of our money, but what can be done about the high domestic price, which we must pay at the pump? That's easy—use an untax to keep the domestic price high. That way, we get the cost increase back in our annual untax refund checks.

Refund checks are great for reducing the pain of high domestic prices, but there's even more help for high prices—that's the double-or-nothing principle. When OPEC has pushed prices high enough, we don't need to up the ante, so the right level of untax is zero—we pay nothing extra. But when we succeed and knock down the world price to a low level, then the untax rate on oil should be roughly twice as high as it is on coal—double. That will keep our oil usage and world prices low.

The OPEC wolf has returned to our door, but our chances are better this time than ever before, for one simple reason: global warming. Carbon caps are now a global phenomenon, and oil is mostly carbon. Carbon taxes are also gaining more acceptance.

As I explain in Chapter 13, the best antidote to OPEC is an international consumers' cartel. The national policy that I discuss in this chapter is less effective, but it provides the basis for the kind of cooperation that a consumers' cartel requires. Part 4 of this book tackles how to organize a cartel.

The New Oil Prices Aren't Like the Old Ones

In 2008, oil prices exceeded their 1980 record by more than a third, but that's only one reason they're more dangerous now. We are up against a new, and likely tougher, opponent.

A $10 rise in the price of a barrel of oil costs U.S. consumers over $70 billion a year. So the increase from $35 a barrel in 2004 is no small matter. But whose fault is it? The Iranian revolution in 1979 cut oil supplies, but as Iran came back on line the Saudis cut back deliberately to keep the price high, cutting their production 75 percent by 1985.

As in 1985, our friends the Saudis are again the only country in the world holding back production to raise prices—but this time they don't have to work as hard. In mid 2008, they're holding back only about 20 percent of what they could produce. That's part of the reason for high prices. But demand has nearly caught up with production capacity. With rapid growth of demand in China and other developing countries, market forces balance supply and demand by raising the price.

Supply has been growing slowly for two reasons. First, the Saudis have long been planning for demand to catch up with the world's production capacity. In 1979, *New York Times* columnist Leonard Silk explained the Saudi policy: "Saudi Arabia has quietly shelved its earlier plans to expand its oil capacity by the mid-1980s to 16 million barrels a day from its present capacity of 11 to 12 million barrels. … They will thus be in a strong position to resist American pressure to expand production, since the extra capacity will simply not be there."

Silk called it correctly. In 2006, Saudi Arabia's production capacity was only 11 million barrels a day, the same figure Silk reported for 1979. It was a smart move. The world did not needed more capacity until about 2003, and since then the shortage of capacity has helped send oil prices through the roof.

OPEC members produced over 30 million barrels a day in 1973 and are producing only a few percent more in 2008. Strangely, the Department of Energy (DOE) has been counting on OPEC to expand its production rapidly as soon as we need it.

Figure 1 shows a 2004 DOE forecast of oil prices, but I have added a line as a reality check. The line shooting almost straight up after 2003 shows what actually happened. The lower lines are all DOE forecasts. Optimistic as they are, though, the DOE's predictions have been getting gloomier for eight years running.

Slowly, the DOE is coming to realize the world is facing a new opponent. Nature is joining OPEC's team. That is the second reason supply is expanding slowly. I'm not saying oil production has peaked, but by most accounts it is getting more difficult and expensive to increase production. After pumping out another 100 billion barrels in the last thirty years, nature is slowing down even the Saudis, and many countries are past their production peaks and in decline.

Between slow capacity expansion and rapid increases in oil demand, it looks like the oil market is going to be tighter in the future than it has been in

Figure 1. The DOE Has Assumed OPEC Would Increase Supply to Meet Demand

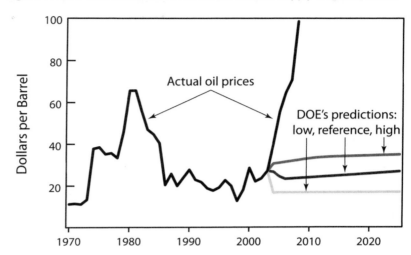

This figure reproduces Figure 26 from the DOE's *International Energy Outlook 2004,* with the addition of the steep line representing actual prices from 2004 through mid-2008. The DOE's high price case proved no match for reality. Prices are for U.S. refinery imports of crude oil in 2002 dollars. These are lower than the spot prices commonly reported.*

the past. With little spare production capacity, we should expect frequent price shocks, if not sustained high prices.

In short, this means the United States should have adopted a serious oil policy long ago. Doing so now will require a shift of political focus that is also long overdue. Fortunately, the same policy will work against both old and new opponents—OPEC and nature.

Pay More to Save Money

OPEC has figured out that to make more money, its members need to sell less oil. That's a hard pill to swallow when they are making $80 to $100 on every barrel they sell. Some in OPEC didn't understand that during Yamani's great energy experiment in the 1970s and 1980s. But OPEC's members may have learned from their early mistakes.

It's time we learned a similar lesson. If we do, we can beat OPEC at its own game. In Chapter 13, I quote Daniel Yergin, author of *The Prize,* explaining that "a tariff in the oil-importing countries ... would transfer revenues from [OPEC's] treasuries back to the treasuries of the consumers." A tariff on oil will transfer revenues from OPEC to the treasuries of the consumers—a tough lesson to understand, but a welcome payoff.

But how does a tariff on oil do that? Can taxing oil really make us richer? It can, and the untax makes it easier to see how this works. When the oil tax is working and has reduced the world price of oil, American consumers are still going to be paying more at the pump, not less, for gas. That high price is necessary to help us fight our oil addiction. But in spite of the high price, gasoline will cost us less—not at the pump, but after the untax refund.

For example, suppose that the price of oil would average $80 for the next twenty years if we did nothing. If instead we put a $40 untax on all oil, we would use less oil. The world demand for oil would decline, and the world price would fall to, say, $70. America would pay less to OPEC and less to domestic oil producers—$10 per barrel less.

But the after-tax domestic price of oil would be $70 a barrel plus the $40 tax—a total of $110 per barrel, way more than the $80 the oil would cost without the tax. But if you've been following this book, you know the trick. With an untax on oil, consumers get back $40 per barrel. It's part of the June refund check. Table 1 summarizes what happens with and without the untax.

Table 1. *How an Untax on Oil Saves Consumers Money*

	With $40 Untax	Without Untax
World Oil Price (Before-Tax Cost of Oil)	$70	$80
After-Tax Cost of Oil	$110	$80
Consumer Refund	$40	$0
Net Consumer Cost	$70	$80

Costs are per barrel of oil.

This is a cautious example, and if the oil price is destined for $200 a barrel without an untax on oil, the savings would likely be much greater. But what really limits the success of this policy is going it alone in a world where other countries, such as China, are increasing their oil consumption. For real success, consumers need a cartel—many consumer nations cooperating—to control the world price of oil.

Here's another interpretation of this example. You've heard of monopolies and monopoly power. That's the kind of power sellers have—that's the power OPEC has. They exercise it by withholding supply and driving the price up. But you've probably never heard of *monopsony* power. That's the economic term for buyers' or consumers' market power. If consumers get organized they can pull exactly the monopoly trick, but in reverse. They can withhold demand—use less oil. As in the early 1980s, cutting back on demand drives the price down.

It's difficult to organize 300 million consumers, which is why we need the untax. Basically it slowly organizes a partial consumer boycott of oil. That's why, as Daniel Yergin explained, "Few things could more quickly arouse the exporters to outrage than the prospect of a tariff in the oil-importing countries." Such a tariff is tantamount to organizing a boycott of OPEC.

In the example, we save only $70 billion a year. But with a higher untax—brought on gradually, or with some help from other consumer nations—a consumers' cartel—we could save much more.

If the Price Is High, Skip the Tax

A carbon tax automatically hits coal hard, because coal has more carbon per unit energy than any other fossil fuel. When it comes to global warming, that's just what we want, because carbon is the main problem. But global warming is not the only reason for a carbon tax. Oil deserves a higher tax than coal for two reasons that don't have to do with global warming. First, oil causes most of our energy security problems, and second, taxing oil drives down the cost of oil.

I will not try to weigh the relative value of these three reasons; the political process will do that. For purposes of discussion, I will simply assume that the extra reasons for taxing oil mean it should be taxed twice as heavily as coal when its price on the world market is low. So if the carbon tax on coal is $30 per ton of carbon dioxide, the tax on oil would be $60 per ton of carbon dioxide.

That works out to about $29 per barrel of oil, or 59 cents a gallon for gasoline. This is a serious incentive, and it is best to work into it gradually. In 2007, such high carbon-tax rates would have collected about $170 billion in revenue from oil and about $280 billion total from all fossil energy. Of course, that is not the net cost to consumers, because an untax refunds all of the tax revenues.

I've discussed two reasons to set the untax rate higher on oil than on coal. But when oil prices are extremely high, it makes more sense to set the tax rate on oil to zero. When OPEC and the world market are essentially taxing oil at a high rate, that is enough. In fact, economic models of global warming policies suggest that a $100 price for oil is as high as the price of oil needs to go until after 2030—assuming we wish to cap carbon at 80 percent below 1990 levels by 2050.

This idea seems pretty obvious. When oil prices go through the roof, we don't need to tax oil. But somehow cap-and-trade advocates always seem to miss it. A national cap-and-trade policy, like one of the many before Congress, will tax oil no matter how high its price on the world market. That just doesn't make sense. But that's how cap and trade works. All industries must pay the same price for carbon permits—and that includes the oil industry.

But using a carbon tax instead of a carbon cap makes it easy to design a policy with a double-high tax on oil when the world oil price is low and no tax on oil when the world price is high. Suppose we consider $100 per barrel a sufficiently high price, but when the world price of oil is low, we think its carbon dioxide emissions should be taxed at $60 per ton. That comes to $29 per barrel. The self-adjusting tax rate should then be either $100 per barrel minus the world oil price or $29 per barrel, whichever is less.

If the world oil price is $100 or more, the tax is zero. If the world price is $90, the tax is $10 and so on down to a world price of $71 and a tax of $29. In that range, the after-tax domestic price of oil stays at $100 a barrel. Below a world price of $71, the tax stays at $29 a barrel. Of course, this plan needs refinement, and variations are possible. My point is simply that a reasonable design is easy to come by.

One necessary refinement is a way to adjust the "sufficient" price, which I set at $100 in my example. This figure should adjust up over time. Moreover, if the actual world price of oil averages, say, $120 for a period of, say, a year, that would indicate that the "sufficient" price should be set at least that high from then on. Such a rule would have the beneficial side-effect of putting pressure on OPEC to prevent such a price shock. There is nothing OPEC members dislike more than to see us increase the tax on oil.

> **Energy Policy #2:**
> **Setting the Oil Tax Rate**
>
> All oil tax revenues should be refunded as part of the carbon untax refund. When world prices are cheap, the oil tax should be higher than the general carbon tax, but the tax rate should fall to zero at some sufficiently high price of oil. That high price should increase a little each year.

Finally, we must face the question of domestic versus foreign oil. Domestic oil companies will argue that we should tax domestic oil at a lower rate than we tax foreign oil. This is identical to giving them an additional markup above OPEC's cartel pricing. If the domestic tax rate is $10 less than the foreign tax rate, domestic producers will set their price $10 above the world price of oil. Some think OPEC's prices are plenty high and that we don't need to give domestic producers an additional markup.

Of course, the oil companies will argue, as they have for the last thirty-five years, that they need more profits so they can find more oil. They will also argue that since their oil is domestic, it does not create energy security problems. There is some truth to both these positions. However, in the late 1970s, when OPEC handed our domestic producers unheard-of profits, Mobil bought the Montgomery Ward chain of department stores, Exxon went into office automation, and Gulf bid for the Ringling Brothers and Barnum and Bailey Circus. Daniel Yergin documents all this in his book *The Prize*. Perhaps the oil companies don't really need even greater profits to find more oil.

Regarding energy security, if we reward the companies for producing more oil now to make us more secure today, that simply means they will produce less oil later, which will make us less secure then. The decision of whether to hand Big Oil extra profits on top of what OPEC hands them will be political, and we should probably ignore the "expert" arguments from oil companies.

～

It is long past time for the United States to wake up and realize that our own oil companies do not control the world price of oil. That era ended in 1973. And we need to recognize that we're the biggest addicts in a world that uses too much oil. All of this means we've handed a lot of power over to OPEC and terrorists. Even so, when they cut supply, our companies still profit. This leaves us with a simple choice. Either we tax ourselves, or we let OPEC tax us, with Big Oil collecting almost half the tax. The difference is that OPEC may forget to mail your refund check in June. And so may Exxon.

The good news is we don't need to tax oil when the world price is astronomical. In that case, the price is already doing a good job of reducing demand. We just need to keep the domestic price from coming down when we succeed in bringing the world price back down. That requires an automatically adjusting tax that goes up as the world price comes down. As that happens, our refund checks grow fatter, although the price at the pump stays high to enforce the consumer boycott of OPEC and Big Oil.

This is not a painful, belt-tightening policy. Spending less on oil and more on other goods will bring down the world price of oil and transfer money from OPEC and Big Oil to American consumers. All this requires is enough foresight and self-discipline to pass a sensible untax. This policy is a triple winner, increasing energy security, slowing climate change, and benefiting consumers. You really can't ask for more.

chapter 20

A Race to Fuel Economy

GM has unveiled cars that on average are nearly a foot shorter and 700 lbs. lighter.

In 1974 the Olds 98 managed only 7.6 m.p.g. on city streets and 11.2 m.p.g. on the highway. In 1977 it posts marks of 16 and 21 m.p.g., respectively.

—*Time* magazine, 1976

"HELL, THE PEOPLE HAVE BEEN TELLING US for years that they wanted smaller, lighter cars," said the vice president of American Motors in 1975. "This industry just has not been listening." But with the oil crisis, people were speaking a little louder. In 1977, the model year before fuel-economy standards went into effect, General Motors raised the average mileage of its fleet by 10 percent in one year.[1] Standards shouldn't get all the credit.

Fuel-economy standards first passed in 1975 when they were set to gradually tighten from 1978 until 1985. From 1985 through 2008, the fuel-economy standard for cars has stayed constant at 27.5 miles per gallon. But the weighted-average fuel economy of cars and light trucks combined has decreased, because most SUVs are classified as trucks, which gives them a lower fuel-economy standard. With a lower standard, the shift to SUVs has brought down the combined average.

In 2006, legislators set the standards to tighten again—starting in 2010. But between 1985 and 2006, with oil prices lower, the automakers had their

---

1. These fuel-economy improvements were actually planned before the oil crisis, but were accelerated by it.

way. As the *Wall Street Journal* explained in 2002, "A national advertising and lobbying campaign led by U.S. auto makers and unions flattened a coalition of safety, environmental and consumer groups—briefly supported by Honda Motor Co.—that had hoped to get the Senate to raise Corporate Average Fuel Economy standards for the first time since 1975."

The fact that automakers and unions "flattened" the standards was a sure sign they were poorly designed. Yes, it also shows the power of automakers, but their power was understood from the start, and a good design would take that into account. Instead, the Corporate Average Fuel Economy (CAFE) standards provoke giants and practically beg to be tampered with.

Imagine a footrace that does not award a prize to the winner. Instead, the race committee, after holding extensive hearings, sets a minimum time for runners. At the hearing, the runners are expert witnesses. They are mad about the cost of the race and have influential friends. For twenty-two years, the runners have said the minimum time was fast enough, and the race committee has accepted it. That's a poor design for a race. But it's a pretty accurate description of CAFE standards, in which the runners—the auto companies—haven't improved for twenty-two years.

In this chapter, I suggest that an old-fashioned race would inspire better performances. If carefully designed, it would also reduce or eliminate the threat to the profits of the Big Three automakers. Best of all, a real race would cut through the red tape that entangles CAFE standards. A race needs no standard at all; it simply relies on competition—just like a market. That's why it beats command-and-control standards.

Getting Rid of Standards

If we stick with standards, how tough should they be by 2020? That all depends on how much is costs for better mileage. So how much does it cost? For any serious level of improvement, no one knows. The automakers may have a rough idea, but they only divulge their most cautious estimates. So how do we end up setting the standard? Cautiously.

There's a better way. It's not particularly new or innovative. Amory Lovins, among others, has been advocating it for years, and it goes by the unlovely name of *feebates*. I like to think of the scheme as a race, and it works like this: Each year, the Environmental Protection Agency (EPA) evaluates the mileage of all the new car models—just as they do now with CAFE standards. Then the EPA hands out prizes. The better the mileage, the bigger the prize—say, an extra dollar for each gallon saved over the life of the car. (The prizes are called rebates in feebate jargon.) The EPA charges the manufacturers of below-average cars comparable fees—in this case, a dollar for each extra gallon used.

This simple system cuts through miles of red tape. The government can reuse it every year, and no car company can testify that it couldn't possibly meet such a tough standard or that it will take an extra three years to meet it. There is no standard. Car companies just do their best. It's a race.

The prize is for gallons saved over the life of the car, but isn't that difficult to estimate? Not really, because it doesn't hurt to use a fairly arbitrary value, say, the lifetime mileage of an average car. Perhaps it's 120,000 miles.[2] The government uses the same value for every car and just divides by the EPA mileage. If the EPA figure is 20 miles per gallon (mpg), the result is 6,000 gallons of gas over the life of the car. This makes the whole scoring process incredibly simple, and it treats every gallon saved the same.

I'll return to the reward per gallon in a moment, but first let's check on who foots the bill. The prizes are proportional to the amount saved relative to the average. Those doing better than average make money, and those doing worse foot the bill. The mathematics of averages guarantees that the total cost of all prizes will exactly equal the penalties. Taxpayers need contribute nothing. The losers pay the winners.

Environmentalists will find several advantages in this system, in addition to the lack of red tape. First, the race can start immediately. If the automakers can't improve the first year, that's OK, but chances are they can. They can produce and sell a few more of their high-mileage cars and a few less of their low-mileage cars. Starting quickly saves years of waiting.

Second, the program motivates all automakers to do better, even those that normally beat the standards by a mile. In a race, the better they do, the more they make. Every carmaker has an incentive to improve.

Finally, the system pushes each carmaker to go as far as it can and to think long term. It is impossible to outrun this nonstandard. No matter how well an automaker does, it can win a bigger prize if it does better.

After considering how to set the prize, I will discuss how to avoid disadvantaging the Big Three automakers. That's crucial, because we want them to stop flattening coalitions of safety, environmental, and consumer groups.

The Size of the Prize

How big should the prize be? The correct answer to that question depends on factors that people almost entirely ignore in the fuel-efficiency debate. To determine what those factors are, I will start with basic principles. What is the purpose of a fuel-economy incentive? Is it to reduce oil use? Yes. But I have

---

2. It does not matter if the estimate is accurate, since a 100,000-mile estimate and a prize of $1.50 per gallon saved is exactly the same as a 150,000-mile estimate and a prize of $1.00 per gallon saved. The prize rate will compensate for any inaccuracy in the estimate.

already concluded that this is best accomplished by an untax on oil, and I'm not taking that back. The main trouble with fossil energy use is the underpricing of fossil fuel, and that is best corrected by correcting the price of fossil fuel with an untax. Underpricing should not be "corrected" with standards or even a fuel-economy race.

The justification for fuel-efficiency standards must be a different market failure, and it is. As I discuss in Chapter 7, car buyers undervalue the future cost of gasoline. Economists call that *consumer myopia*—nearsightedness. Because consumers ignore part of future gasoline costs, they don't spend enough on improved fuel efficiency. So it doesn't pay carmakers enough to make high-mileage cars.

Two other reasons for a fuel-efficiency standard are even more difficult to place a value on, and economists note them less frequently. Still, they may play a significant role. First is the status value of a bigger, more powerful car. Outside the economics world, this effect is widely recognized—for example, in the Volkswagen commercials claiming the Passat has the lowest ego emissions of any German-engineered car. Unfortunately, the status value of low ego emissions seems to have been short-lived, and VW quickly returned to claiming that the Passat 2.0 Turbo beats some models of BMW, Lexus, and Mercedes in going from 0 to 60 miles per hour.

So what's wrong with status? Well, nothing. But a race for status produces a classic economic inefficiency. Status is a relative measure. When one person gains in status, others automatically lose status relative to the gainer. If everyone bought a slightly smaller and less powerful car, it would not change anyone's relative status, and everyone would save a little money. For this reason, a status race based on size and power is a waste of money, and the fuel-efficiency race works against this waste.

The third market failure is the safety race—or the SUV arms race, as it has been called. So what's wrong with safety? Well, nothing. Yes, this one works exactly the same way as the status race. The source of the problem is that when I buy a bigger car, I'm better off—safer—but others are less safe. Few consumers take into account the safety of society as a whole when they buy cars. Instead, they ignore the fact that increased car weight has a negative side effect on other drivers, and they buy cars that are heavier than what is best for society. We would all be safer if we spent less on weight and more on safety features that don't make others less safe. A fuel-economy race helps correct this problem.

Now, I do not want to give the impression that the feebate approach can be fine-tuned to solve these problems. Rather, I want to question the extreme neoclassical economics position, which holds that the car market is optimal and that the only energy market failure is the price of carbon. (See the opening quote of Chapter 5.) In fact, the car market is suboptimal for several reasons, and the most obvious three all suggest that an incentive for increased fuel efficiency in cars would improve the market. So fuel-economy standards, or the prize in a race to fuel-economy, should not be set to correct the low price of carbon. These should be set to correct problems with the car market itself, which work to decrease fuel economy relative to what's best for society.

Returning to my initial question of how big the prize should be, let's check with the National Highway Traffic Safety Administration (NHTSA), which sets and administers CAFE standards. When considering where to set the standard, the NHTSA takes the view that only improvements that save enough gas to pay for themselves in the first four and a half years of driving should be mandated. The NHTSA's justification is that people own cars only 4.5 years on average and ignore the benefit of fuel savings after that. Orthodox neoclassical economists will argue that people don't ignore those savings but instead factor them into resale value. However, that's missing the point. The NHTSA finds significant fuel-economy potential even though it considers only 4.5 years of fuel savings. This indicates that consumers and carmakers are taking into account less than four and a half years of fuel savings. Otherwise, the NHTSA would not find such cheap fuel-economy measures still unimplemented.

Moreover, those savings would come on top of fuel-economy measures already induced by present CAFE standards. In short, the NHTSA's implicit position is that consumers ignore more than half the future cost of gasoline at the time of car purchase. That's why there's room for standards that actually save money. And since standards are justified on the basis of fuel saved during only the first 4.5 years, the NHTSA, as well as consumers, is ignoring fuel savings during the final six-plus years of a car's life.*

For the sake of discussion, suppose consumers do ignore half of future fuel costs and suppose gasoline costs $3 a gallon—including the untax. Then the prize should be set to $1.50 per gallon.

At this prize rate, if the NHTSA's designated average car life is 120,000 miles, each 30-mpg car manufactured will bring in $3,000 more prize money than each 20-mpg car manufactured. That's a strong incentive for car companies to improve.[3]

How expensive is this? On average, the prizes cost car companies nothing. Making better cars is costly, and the consumers will pay that cost. But if consumers are as nearsighted as experts assume they are, the cost of the improvement will be less than the cost of the gas saved, so consumers come out ahead.

If, contrary to what the NHTSA assumes, consumers already take full account of gasoline savings, and if they don't value status, and if they fully consider the safety of others, then no net savings would result from increased mileage. If the car world is already perfect, as orthodox neoclassical economists

---

3. This is because a 20-mpg car will use 2,000 gallons more than a 30-mpg car, and 2,000 gallons at $1.50 per gallon-saved comes to $3,000.

assume, no policy can make it better, and any change will make it a bit worse. Betting on a race to fuel economy is taking a small risk—that we will disturb a perfect market—for a large potential gain.

Helping the Big Three

The Big Three—General Motors, Ford, and Chrysler—may not be the best carmakers in the world, but they're our carmakers, and they're getting better. Moreover, losing the Big Three would disrupt tens of thousands of lives. We can lament their role in the advertising that helps form the public taste for gas-guzzlers and encourages status seeking, but consumers are far from blameless. I'd like to suggest that fixing the energy problem is more useful than bashing the car companies, and at the moment it appears that we have an opportunity to fix the problem.

Let us bury old grudges and get down to business. It is not particularly helpful when U.S. automakers and unions flatten coalitions of safety, environmental, and consumer groups. As long as the automakers feel threatened, they are likely to continue this tradition. So let's look for a way to induce them to cooperate and like it—or at least not hate it.

I believe their main fear, particularly with high oil prices, is competition from the likes of Toyota and Honda. It may seem that any race for fuel efficiency is just going to favor those competitors to the detriment of the Big Three. But that conclusion is premature. Remember, we are designing this race. My suggestion is to base prizes on improvement rather than absolute achievement. In such an improvement race, the C student may well have the advantage over the A student. Merely improving to B-minus is more progress than the A student can hope to make.

And moving from a C to a B-minus helps conserve more fuel than you might expect. Improving a car from 100 to a million mpg doesn't save much gas, because a 100-mpg car doesn't use much. The same principle holds in more realistic realms. Improving a car from 40 to 70 mpg saves less gasoline than improving one from 15 to 18 mpg. If improvement is the goal, starting with gas-guzzlers is a big advantage. Now, who might that favor? And it's good for energy security and the climate, because improving gas-guzzlers does the most good.

How do we redesign the race so it rewards improved fuel economy? Only a small tweak is needed. We use the year before the race was proposed as the base year. (This avoids giving car companies an incentive to sell extra gas-guzzlers in the base year to give themselves an easy starting point.) Using standard EPA data, we calculate the average gallons per vehicle for each car company in the base year. That's its base usage. The company's prizes are based on how much it beats its own base usage. The formula is given in "Details of the

Race to Improve," and it shows that if none of the companies made any progress in the first year, they would all win zero. Toyota will have no advantage over GM to start with, and GM will have the advantage of more room for improvement. The Big Three should love this race.

～

We wish to use less oil in order to reduce its price on the world market, improve energy security, and reduce climate change. To reach all those goals, we don't need to adjust the auto market; we simply need to raise the price of oil with an untax. However, problems in the market for cars and light trucks do justify incentives for automakers. The key problem lies with consumers who ignore part of the future cost of gasoline when they purchase a car. Experts most familiar with CAFE standards suggest that consumers undervalue fuel costs by roughly half.

CAFE standards have resulted in remarkably little progress over their thirty-three-year history, mainly because powerful interests oppose them, and the use of standards facilitates delays, backsliding, and general obstructionism.

A simple race design would prove more effective and also more difficult to obstruct. Basing rewards on improvement instead of achievement should favor the Big Three instead of threatening them, and it will not reduce their incentive to perform. If expensive hybrids prove to be the best design, building these could well increase the profits of the auto industry at the expense of the oil industry, as consumers spend more of their money on cars and less on oil.

Details of the Race to Improve

Racing to improve, instead of racing to win, could cause trouble if the rules are not carefully designed.

Since the race holds different car companies to different standards, some might be encouraged to specialize in gas-guzzlers or sell a factory that makes efficient cars to a company with a low standard. It's important that the race not have such side effects. It should do its job and leave other aspects of the industry alone.

The following formula represents the proposed improvement race:

$Score = N \times (NAG - G) +$ handicap,

where, N is the number of cars a company makes, NAG is the national average number of gallons used per car, and G is the average number of gallons per car for a specific car company. Except for the handicap, this is exactly like the original race design that judges all by the same standard.

If N did not change, the handicap would be fixed from the start at $N \times (B - BNAG)$, where B is a car-maker's base-year average gallons/car and $BNAG$ is the base-year national average. But since N can change, the handicap could cause trouble unless we cap N at its starting value when calculating the handicap. This means increases in car production will not receive the benefit of the handicap and must compete with the current national average.

A simple adjustment is still required to make the race self-funding. The prize equals the *Score* times the prize rate of, say, $1.50 per gallon.

chapter 21

Crash Programs

This is still the only country where people say with a straight face that anything is possible—and really believe it.

—Senator Lamar Alexander, 2008

WITH PRESENT TECHNOLOGY, we could save a lot of fossil fuel at little cost. But eventually the world must switch to energy sources that have not yet been invented, and the sooner the better. So what will speed the invention of new technology? Some say only the government, and others say only the private sector.

The advocates of a crash government research program point to the Apollo program, which put a man on the moon in July 1969, or the Manhattan Project, which produced the two atom bombs dropped on Japan in 1945. On the other hand, venture capitalists and economists advocate private-sector research, though neither group is opposed to a role for government. They just don't want to put all the research eggs in the government's basket.

The fourth part of my Core National Energy Plan is a proposal that the government ramp up its research program to at least ten times its present level. But the government should spend the money on research, not production subsidies. And we should not bet the farm on breakthrough technologies produced by crash government programs.

Figure 1. Federal Energy Research Has Fallen to 1/100 of 1% of GDP

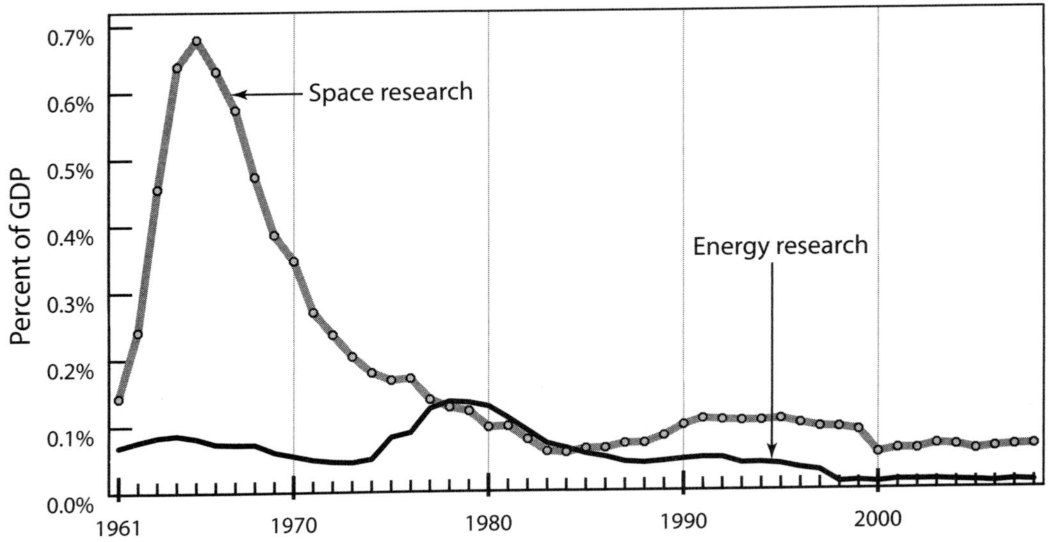

The Apollo program dwarfs the energy research that occurred during the OPEC crisis. Space research still exceeds energy research several times over. As a percentage of GDP, energy research is far less than before the first OPEC crisis. Data from the National Science Foundation.*

The Government Drops the Ball

The federal budget for energy research, including nuclear energy and fossil fuel research, is 1 percent of 1 percent of our gross domestic product (GDP). That's one ten-thousandth of GDP. In 2007, we spent six times that much exploring outer space.

As I write this in 2008, we are spending 340 times more importing foreign oil than on figuring out how not to. Compare the great mountain on the left of Figure 1, which represents money spent on the Apollo program as a percentage of GDP, with the barely visible line at the bottom right of the graph, which represents today's federal energy research budget. One percent of 1 percent is about how much richer the country gets every day. If we permanently doubled federal expenditure on research, it would cost the same as delaying the country's economic growth for one day.

President Bill Clinton ramped down spending on energy research to its present low rate while the price of oil was low. That's no excuse, but it probably explains quite a bit. But continuing at such a low level while oil prices rose from $30 a barrel in January 2001 to over $100 a barrel in 2008—what explains that?

Venture Capital

Big companies like General Motors quietly fund their own internal research, but little companies need venture capital, and they often like to brag about their projects. One story of venture capital and energy innovation holds particular interest—the story of the race between lithium-ion batteries and hydrogen fuel cells.

General Motors and Toyota would both love to make a practical electric car. To hedge their bets, they've been looking into both battery cars and fuel-cell cars. Batteries need to be recharged with electricity every night, while fuel cells make their own electricity from hydrogen. So fuel-cell cars need hydrogen filling stations and super-high-pressure "gas tanks" for hydrogen.

Batteries are expensive, as are fuel cells and hydrogen tanks. So engineers and scientists have been racing to bring down the cost. General Motors and Toyota have been waiting to see who wins. Starting in about 1999, Amory Lovins and the Ballard fuel-cell company claimed fuel cells were about to break the price barrier. The nuclear industry, always a fan of the hydrogen economy, jumped on the bandwagon. The government finally joined the party in 2002, when it launched the FreedomCAR Partnership.

By then the hydrogen bubble had burst, and Ballard's stock had fallen to $52 from a high of $93. But governments are slow to catch on, so a year later, with Ballard's stock down to $18, President Bush announced the $1.2 billion Hydrogen Fuel Initiative in his 2003 State of the Union address. Ballard's stock price continued its slide into the low single digits. By 2008, federal government spending was up over $300 million a year for hydrogen cars.

In fall 2001, just as the hydrogen craze was peaking, an MIT professor named Yet-Ming Chiang and two partners founded A123 Systems in Watertown, Massachusetts. They quickly raised about $12 million for what proved to be the hopeless idea of making chemically self-assembling batteries. But in 2002, Chiang discovered a way to make lithium-ion batteries release their power more quickly, which can improve a car's acceleration. The company's KillaCycle, an electric motorcycle, is now the fastest electric vehicle in the world in the standing-start quarter mile, sprinting it in under eight seconds.

In 2003, A123 met with representatives of Black and Decker, who wanted better power-tool batteries, and by 2006 Black and Decker's customers were buying power tools sporting A123 batteries. DeWalt's best power tools soon followed suit. The new design makes the batteries quicker to charge, less likely to catch fire, and longer lasting. A123 has now raised about $150 million in capital, and General Motors is planning to use A123 batteries in the Volt, its plug-in hybrid scheduled for production in 2010.

Now, the federal government can spend a lot more than $150 million, and it has thrown about ten times more than that into hydrogen cars. But in

2008, the *Wall Street Journal* quoted General Motors' vice chairman Bob Lutz as saying, "If we get lithium-ion to 300 miles, then you need to ask yourself, Why do you need fuel cells? We are nowhere [near] where we need to be on the [fuel-cell] costs curve." So it looks like little A123 may win the race against the federal FreedomCAR Partnership, which started backing hydrogen just as the bubble was bursting. General Motors has picked its batteries over government-backed hydrogen fuel cells.

Why is A123 winning? Some luck was involved, but A123's success also illustrates an important principle. I'll let venture capitalist Vinod Khosla explain. He's a cofounder of Sun Microsystems and a major funder of Google and Amazon.com. He has degrees in electrical and biomedical engineering as well as an MBA from Stanford. Says Khosla, "Taking a risk with your own money, which venture capitalists do, is much better talked through than the risks that the government will take. They [government officials] listen to pundits and pontificate without understanding the technology."

Notice a key phrase: "with your own money." The Department of Energy (DOE) is spending your money, and Khosla is spending his money. He says the "risk … is much better talked through" when he's spending his money. Having spent several years at a national laboratory funded by the DOE, I have to agree. While I always found that the DOE's employees were trying to spend our money wisely, I was never impressed by their ability to do so. In fact, as Khosla suggests, the DOE discussions are not always well informed technically, and the DOE's perspective on risk and payback is none too clear.

This shows up in the race between hydrogen cars and battery cars. Government officials made their decision in favor of hydrogen cars after listening to pundits, just as Khosla would expect. Unfortunately, pundits get revved up only when a craze is in full swing. By the time the government catches on, it's usually too late. Of course, venture capitalists make similar mistakes, but the best capitalists make quicker, more thoughtful decisions. To see how hard it is to get the government to change course, consider the efforts of the National Academy of Sciences (NAS).

When the government brought in the NAS to review the federal hydrogen program in 2004, the NAS warned the DOE to "keep a balanced portfolio of R&D efforts," because "if battery technology improved dramatically, for example, all-electric vehicles might become the preferred alternative." In a second review, published in 2008, the NAS remarked, "There seems to be a lack of urgency in finalizing and executing the R&D plan for plug-in hybrid vehicles [battery-powered cars]." The DOE had been paying attention to politics, not to the experts it had asked for advice.

Khosla's remarks reflect a consensus within both the private-investment community and economics: When it's your money that's at risk, you think harder about how to spend it than when you are just part of a bureaucracy, spending taxpayers' money. Not only that, if you don't think hard, you soon lose your money. So venture capitalists like Khosla who remain in the business and have the most to invest tend to be thoughtful as well as talented.

A Role for Government

The market is best at picking technologies that have been invented but not commercialized. It is also best at inventing new technologies that are not too futuristic and not wildly expensive. But when research is too expensive or too fundamental, it doesn't pay off for those doing the research—even though its social benefit may be great. In these cases, the market drops the ball.

For example, present-day nuclear reactors are based on A-bomb technology, but future nuclear reactors may be fusion reactors, which work a bit like hydrogen bombs. Fusion reactors require "heavy hydrogen" for fuel. Only 1 in 3,000 hydrogen atoms is heavy, but this fuel is so powerful, and the oceans contain so much hydrogen, that we would never run out of energy. Moreover, fusion reactors are cleaner than standard reactors, and nothing in them is critical for making bombs.

But fusion reactors are so expensive and will take so many decades to commercialize that the private market would not touch them for another hundred years. Fusion research needs government funding.

As usual, the federal government guessed wrong and started pouring money into fusion reactors in the late 1950s. This was likely way too early, but the effort has been internationalized, and the current $10 billion test reactor should be operating in southern France by 2016. Ten billion dollars is peanuts compared with a decent chance of solving the world's energy problems—even if it takes fifty more years of development.

So an exploration of fusion is one example of research that governments need to fund. Another example is clean coal technology, including hydrogen production with 90 percent carbon capture—the system FutureGen was to use before the government canceled the project.

Basic energy research is an appropriate arena for public funding. Many good ideas are too advanced for venture capitalists to fund. Quite likely, ten times the current level of funding would make sense. That would be about $15 billion a year—still a tiny fraction of spending on the Iraq war. Greater levels might pay off as well. But the increase in spending should be accompanied by greater input from the scientific community.

Atom Bombs and Lunar Landings

Senator Lamar Alexander of Tennessee, who chaired the Senate Republican Conference in 2008, proposed that the United States launch a new, five-year "Manhattan Project" to put America "firmly on the path to clean energy independence." I applaud him for his goal, to enhance "clean air, climate change, and national security."

Alexander explains that his energy project is aimed not at "any of the things that we already know how to do," but at "real scientific breakthroughs." In fact, nuclear fusion and FutureGen are two of his seven "grand challenges" on the path to energy independence. Both are appropriate.

In my view, nothing is wrong with the concept of a Manhattan Project for real scientific energy breakthroughs. That said, I believe the Manhattan Project/Apollo rhetoric causes confusion. A crash program to make real scientific energy breakthroughs makes sense—provided we avoid two errors:

- ▶ Using a crash program when the market would work better.
- ▶ Nixing other approaches in hope of high-tech miracles.

These errors are similar to one found at the heart of many energy policies— the tendency to think we're the Lone Ranger, in possession of a silver bullet. Economists think the price of carbon is a silver bullet. Amory Lovins thinks it's his Hypercar. To physicists, it's energy efficiency. Others say solar, wind, or maybe hydrogen.

The Manhattan Project/Apollo approach follows in this tradition, with the "crash program" itself as the silver bullet. Do you want solar power to work? Use a crash program. Fusion power? Set up a Manhattan Project. Wind, plug-in hybrids, building insulation? An Apollo program is the way to go.

Manhattan Project/Apollo advocates suggest that the energy problem is like the challenge of building an A-bomb or landing on the moon. Or per-haps it's like seven A-bomb problems. So the federal government should put together a team, or seven teams, preferably with "centralized gruff leadership," as Alexander suggests. Then throw in some money and a deadline—as with the Apollo program—and bake for five years.*

This can work—on the right problem. But the broad goals of climate stability and energy security are not much like building an A-bomb or landing on the moon. Those two projects had four important characteristics:

- ▶ They were narrowly focused—build two bombs or land one man on the moon and bring him back alive.
- ▶ The projects did things the market could not possibly do.
- ▶ They were purely technical challenges, with no social changes required.
- ▶ Experts had judged that the projects were technically feasible.

Solutions to our energy problems share none of these characteristics. Today's energy challenges are more akin to fighting crime or ending poverty. They are broad and multifaceted and involve a lot of human behavior.

I could be wrong. A Manhattan Project or two just might fix the entire energy problem. A crash program could invent the ultimate, dirt-cheap solar cells and dirt-cheap superbatteries. Those together could fix everything. I have no objection to trying for a miracle—but I don't want to bet the world on it, or even our energy security.

Green Crash Programs

The crash program idea has environmentalist champions as well as Republican champions. This is not about politics. Ted Nordhaus (not to be confused with economist William Nordhaus) and Michael Shellenberger, two longtime environmental pollsters, have been arguing for an Apollo-type energy program for several years. They've written a book, *Break Through,* and started the Breakthrough Institute, and they're pushing for $30 to $80 billion a year to fund a program that would bring forth "dramatic technology breakthroughs to bring down the price of clean alternatives."

There's nothing wrong with hoping the government will make dramatic technology breakthroughs, and perhaps it's worth spending as much money as they suggest. It's less than what we spend on military research. But what worries me is the pair's rationale.

Writing in the *Harvard Law and Policy Review,* they explain their theory: "The transition to the clean energy economy, … like the previous technology revolutions, will require substantial public investment [in order] to occur quickly and completely." Is that right? Previous technology revolutions required substantial public investment? They list three examples: the Internet, personal computing, and the transition to the petroleum economy. Were these Apollo-style programs?

In 1969, the Internet was invented as a military project. Was this part of an Apollo program to invent the World Wide Web? Far from it. Military research has been like Apollo on steroids for the last fifty years. Every once in a while, something like the Internet spins off by chance from some little side effort. Surely, this hyperexpensive, just-by-accident approach can't be what Nordhaus and Shellenberger have in mind.

Personal computers? Steve Jobs and Steve Wozniak may be surprised to hear their garage was a secret multibillion-dollar crash program. The transition to a petroleum economy? John D. Rockefeller must be turning over in his grave trying to recall the federal crash program that built his oil empire.

And then, lest you forget, there was the Wright brothers' Apollo program, Marconi's Apollo program to invent the radio, Alexander Graham Bell's Apollo telephone project, and Edison's Apollo lightbulb program. But Nikola Tesla may be the most upset of all. He was literally digging ditches in New York City to fund his research into the AC electric motors that Westinghouse used to usher in the modern age of electricity. If only he had known of his billions in federal funding.

Next, Nordhaus and Shellenberger inform us that "energy is the least innovative sector of the economy. Coal and oil have been in widespread use for the last 200 and 100 years respectively." That makes as much sense as saying: Electronics is the least innovative sector because electricity has been in widespread use for 140 years. But they have their reasons for believing that energy is the least innovative sector: "First is that national electricity grids are tailored for large centralized plants." But large centralized computer chip plants have allowed incredible innovations in computers, and the prejudice against central power plants is a back-to-nature fantasy. In fact, the national electricity grid continues to facilitate all the most promising advances in clean electricity generation—wind, clean coal, and the most cost-effective solar. Nordhaus and Shellenberger have got to be kidding—except apparently they're not.

No innovation in energy? What about hybrid cars, rapid advances in wind technology, the new solar thermal installation in Arizona, the A123 battery company, compact fluorescent bulbs, and the even newer LEDs? The first U.S. cellulosic ethanol plant opened in 2008. But if you want to be amazed by American ingenuity, go to www.technologyreview.com/energy. This is MIT's technology magazine, and it covers five areas of technical innovation. Energy is one of the big five, and the magazine reports on a seemingly endless list of new research breakthroughs and new energy products.

Nordhaus and Shellenberger are pollsters, not economists (see "Crash Program Polling"), so it may be unfair to criticize their economics. But they seem to enjoy their reputation as the bad boys of environmentalism, and apparently they think they've discovered that one of the most basic market principles is wrong. They think markets don't work well for stimulating innovation. Economists, they say, advocate carbon pricing to change buying habits but have forgotten about innovation and investment.

Market economies, they tell us, always "require substantial public investment [for technology revolutions] to occur quickly and completely." But this would mean that the technology revolutions that brought us the cell phone, the personal computer, and the Internet were sluggish and incomplete. True, a few people don't have access to those new technologies. But that problem is

---

Crash Program Polling

Pollsters Ted Nordhaus and Michael Shellenberger set out to prove that their crash energy program is popular. They asked people to react to the following two proposals:

1. "Make energy sources that pollute the atmosphere *cost more*. Polluting energy sources include gasoline, home heating oil, and coal-burning power plants."

2. "Make clean energy sources *cost less*. This includes solar and wind energy."

Naturally, 92 percent of those polled favor making clean energy cost less, while only 46 percent favor making polluting energy cost more.

Nordhaus and Shellenberger neglected to mention to those polled the taxes needed to pay for their crash program—up to $1,000 per year for a family of four. And when they asked about carbon taxes, they did not mention that these could be 100 percent refunded and would still work.*

not a breakthrough problem; rather, it's a matter of helping the poorest or most remote catch up with standard technology.

But perhaps Nordhaus and Shellenberger are right when it comes to energy, "the least innovative sector of the economy." Perhaps markets stimulate other innovations, but not energy innovation. Perhaps, as our pollsters say, "Investors are often reluctant to expend their revenue on risky, innovative, and costly ventures without government regulation."

But if investors are so timid about energy, how do we explain the constant comparisons between the current green-tech boom and the dot-com boom? Were dot-com investors risk-averse and shy of innovation? In June 2008, the U.K.'s *Guardian* reported that "more than $3.6 billion of investment cash poured into the [green-tech] industry in the US alone, on top of $2.9 billion in 2006, in itself a 78 percent jump from the previous year."

In May 2008, the *Boston Globe* reported, "Many of the North American and European investors who sunk $5.2 billion into 'cleantech' companies last year—up [629 percent] from $714 million in 2001, according to Cleantech Group LLC—are alumni of the last high-tech boom and bust." The investment community agrees on two things: Investments in clean energy technology have been booming, and it's risky business. That's exactly the opposite of the Nordhaus-Shellenberger theory of economics.

So why is this risky, innovative investment boom happening? The cost of "dirty energy" is going up—the factor that Nordhaus and Shellenberger say doesn't matter. In reality, investors do care about current higher oil prices and expected price increases due to carbon pricing policies Nordhaus and Shellenberger say won't work. The cost of dirty energy—the one factor Nordhaus and Shellenberger dismiss—is the one factor driving the clean energy investment boom.*

Crash Programs versus the Market

There need be no conflict between those who advocate public funding of technology innovation and those who favor private funding. Crash programs are for things the market is not good at, but the market is good at plenty of inventing and investing. The key is not to bet too much on one or the other—crash programs or the market—but to look carefully and figure out which is good for what.

Senator Alexander's first grand challenge, to "make plug-in hybrid vehicles commonplace," provides an example of overselling the Manhattan Project concept. As we've seen, the market is now doing a great job with plug-in hybrids.

But what if we want to double the speed of progress on plug-in hybrids? Should we start a crash government program? There's a faster approach. The carmakers are already in the business; they're pros. They should implement the crash program, and they will if we motivate them.

With the race to fuel economy that I described in the last chapter, motivating them would be trivial. Just increase the prize from $1.50 per gallon saved to, say, $3.00. Remember, this is 100 percent self-funding, so it requires no new taxes. But even $1.50 per gallon would probably be enough; in fact, the car companies already seem to be engaged in fairly intense competition for the plug-in hybrid market. With prizes funded by the laggards, everyone is fully motivated, and there's no need for billions of federal dollars.

But I also agree with Senator Alexander. The federal government can play a useful role in funding advanced battery research. It's worth looking ten years ahead, and venture capitalists do a poor job of that.

Nordhaus and Shellenberger favor solar power over plug-in cars, wanting to take advantage of the supposed empirical law that doubling solar production capacity reduces production costs 20 percent. Their goal is to increase sales by subsidizing solar. This is the opposite of Alexander's philosophy of funding research that could lead to "real scientific breakthroughs." Nordhaus and Shellenberger would be more convincing if they stuck by their favorite term—*breakthrough*—and forg0t massive subsidies for existing technology.

The terrible thing about the government subsidizing large-scale production to bring down the price is that, when a breakthrough comes, all the factories that were built to take advantage of large-scale production will become obsolete. The government should instead fund advanced solar research. Then, when the breakthrough happens, the technology will take off on its own. That day will come much sooner if dirty energy costs more.

⁓

Clean energy research funded by the federal government has reached the absurdly low level of one ten-thousandth of the GDP. Given that this is 340 times less than we pay for imported oil in 2008, not to mention climate problems, we should immediately begin ramping up to at least ten times this level—a bit more than we spend on space research.

Markets under-reward advanced research—the kind that leads to major breakthroughs—and this flaw justifies government spending. But it does not justify spending on things the market does well. In particular, it does not justify subsidies that partially and erratically replace proper carbon pricing.

A distrust of carbon pricing and a desire for silver bullets appear to motivate the crash program approach to government spending. But the distrust of carbon pricing is based on the great cost confusion, as I explain in the next chapter.

chapter 22

The Great Cost Confusion

Opponents say it's going to cost so much money to address. And I say, well, hell, go ahead and spend it.

—Oilman T. Boone Pickens on climate change, 2008

NEVER HAS CIVILIZATION SWITCHED ENERGY sources as quickly as we may need to now. Never has American energy policy made much difference on balance, with pluses roughly canceling minuses. Success requires a powerful and coherent policy. Of the many obstacles blocking the adoption of such a policy, two loom largest: subsidy politics and the great cost confusion. Both distort the roles of the government and the private sector in ways that lead to wasteful subsidies and, ultimately, failure.

Subsidy politics is a dance of meddlers. The government uses subsidies to meddle in the market, trying to pick and foster winners. Industry lobbies the government for subsidies and, in doing so, meddles with policy.

The great cost confusion makes it easier for lobbyists to coax subsidies out of government. The confusion occurs when policy makers assume carbon pricing revenues are net costs to the nation. Having assumed this, they feel free to spend the revenues, and the lobbyists are ready with suggestions. The meddlers dance to the tunes of the great cost confusion.

The government's proper role is not to subsidize technologies but to identify market failures and fix them with minimal intervention. This approach takes full advantage of the market's power and maximizes the government's impact. The proper role of the private sector is to respond to market forces, not to lobby for subsidies.

Subsidies: Fuel for Meddlers

Markets cannot do certain jobs. For example, the market cannot, on its own, evaluate and take into account the cost of military expenditures to protect oil trade routes. So the government should intervene to fix this problem.

Suppose the international chocolate trade, like oil, needed government protection, and the protection cost was not factored into the price of chocolate. Free protection is a subsidy to chocolate, and it keeps its price too low. With the price too low, people eat too much chocolate, and the government spends too much money protecting global chocolate routes. After a bit of lobbying by other candy makers, the government might decide to subsidize peanut brittle and licorice to reduce chocolate consumption and save on the military costs of protecting chocolate.

This would "work." But it's a distraction. The government has ignored the real problem, the chocolate subsidy, and focused instead on inventing new subsidies to distract people from the cheap, subsidized chocolate. Peanut brittle is like ethanol, and licorice is like synfuels. Soon, they will be subsidizing candy canes (solar roofs) and who knows what. But just make chocolate pay its own way, and people will switch to other candy exactly to the extent they should. Don't let lobbyists design balanced subsidies—they can't and they won't even try.

Meddling by the government usually takes the form of subsidies. But instead of meddling the government should follow the fossil philosophy I described in Chapter 1 and treat the problem, not the symptom. An untax on chocolate or oil would do just that and completely eliminate the need for subsidies.

Some proponents of subsidies dislike them but think they are a necessary evil. Business, they think, will never do the right thing without a bribe in the form of a subsidy. I would like to offer another perspective. Energy policy is just too big for this approach to work. The subsidies would be too big to hide, and once visible to the public, would discredit any energy plan. Of course the main reason to avoid most subsidies should still be that, in practice, they are largely a waste of money.

Too Big to Hide. Moving away from fossil fuel will shift hundreds of billions of dollars from oil, gas, and coal to high-tech cars, super-high-tech coal plants, and the like. You can't hide any policy that affects what happens to that much money.

Because a policy that drives a substantial shift to cleaner energy sources cannot be hidden, it needs to be cost-effective so people feel they are getting their money's worth. It needs to avoid even the appearance of corruption. Letting the market pick the winners avoids the appearance of favoritism that would otherwise make any large-scale policy unpopular. Even more important, the market, which is you and I and businessmen, will do far better at finding the cheapest way to avoid high fossil-fuel prices than will special interests.

Special Interests Bring Discredit. Most energy schemes rely on support from one or more political interest groups on the theory that what's needed most is political muscle. But such support can be generated from two types of benefits—targeted benefits and societal benefits. The real strength of good energy policy is the societal benefits it confers.

Almost every energy program—even those backing the worst synfuels—attempts to harness the politics of targeted benefits. Ethanol helps farmers. More coal helps coal miners. Blocking fuel-economy standards helps autoworkers. Solar panels and wind turbines have their own constituencies. Such targeted benefits bring some political support, but at great cost. Targeted benefits interfere with the perception of a broad social benefit.

Another broad policy, the war on terrorism, is sold on the basis of broad social benefits—safety for everyone. What if it were sold on the basis that it helps the oil industry, benefits Halliburton's employees, and stimulates certain manufacturing sectors? All this is true. But arguing for such targeted benefits would delegitimize the social benefit argument for the policy. The same is true for energy policy.

But targeted benefits pose another, even greater, risk. Every target group has its lobbyists. They meddle with policy so the government will meddle favorably in their part of the market. The track record of such meddling is abysmal. It undermines energy policy by making it ineffective and bringing discredit upon it.

The Great Cost Confusion

Government and special-interest meddlers are aided and abetted by one more villain: the great cost confusion. The great cost confusion comes from mistaking dollars transferred for actual costs—that is, from assuming that carbon revenues are a national cost without first checking to see where they are going.

Economists distinguish between national costs, which they call social costs, and transfer payments. Transfer payments transfer money from one person to another, so there is no net cost to the nation—no social cost. If you buy a popular book, it may cost you more than the

More Jobs and Fewer Jobs

Every program that spends money claims to "create jobs." In reality, when one job is created, another is lost, except temporarily in a recession.

If Iowa makes ethanol and sells it to New York, the money that flows into Iowa flows out of New York. Iowa gains and New York loses jobs. Money that New Yorkers spend on ethanol, they do not spent on other goods and services, for example, eating out.

The awful truth is that even if ethanol replaced OPEC gasoline at no extra cost, the money not spent importing oil would mean less money spent on U.S. exports, and there would still be no reason to bet on a net increase in national employment.

So job creation does not help the country as a whole, unless the new jobs are more satisfying than the jobs that are lost. But that's hard to prove without knowing what jobs are being lost.

true cost of the book. The extra payment is profit for the publisher and author. It's a cost to you but a gain for them. So it is not a national or social cost, it's a transfer payment. However, a law that requires everyone to chip in $20 for fireworks, which are shot off on the Fourth, imposes a social cost. It also provides a benefit—a beautiful display—but the cost is undeniable.

To put these concepts in context, carbon-tax revenues can be recycled through refunds—the untax—in which case they are transfer payments and not a social cost. Or the revenues can be spent on fireworks or clean energy projects, in which case the revenues become a social cost, and that cost equals the revenues spent. Yes, there will be a benefit, but the cost is undeniable.

This may seem too obvious to be worth the fuss I'm making, but collect the revenues with a carbon tax, and even experts with PhDs can't figure out they should check where the money's going—or at least many of them still haven't. Most economists say carbon-tax revenues should be used to reduce some tax you would otherwise have to pay. That's a transfer payment. We pay carbon taxes, but others pay less of some other tax so the tax does not make us poorer on average—there is no cost on average.

The untax doesn't reduce taxes; it gives the money back directly with a check in the mail. That's an even more obvious transfer payment. No, it doesn't put it all back exactly where it came from—that would be going in circles. Some people—those who use less carbon—get more back than they paid, and others get less back. But, on average, there is no social cost at all. Understanding that the net social cost of these money flows is zero matters a great deal, because those who think that revenues are automatically a social cost feel no obligation to return the money the government has taken. They start thinking carbon-tax revenues are free money, and they dream up things to spend it on. That turns the revenue into a true cost to society, to the nation as a whole—but it doesn't have to be like that.

It might seem impossible that experts, politicians, and almost everyone else discussing energy policy still think they can figure out costs without checking to see what the government does with the money it collects. But this error is so common that while I was writing this chapter, my wife just happened to send me an article that illustrates the great cost confusion perfectly.

The piece was an op-ed by Monica Prasad in the *New York Times* titled "On Carbon, Tax and Don't Spend."* In it, she notes that "carbon tax discussions always seem to devolve into gleeful suggestions for ways to spend the revenue." That's exactly my point. But in the end, what does she recommend? "We should follow Denmark's example … and return the revenue to industry through subsidies." Subsidies! After tweaking everyone else for gleefully spending the money, she suggests spending it on subsidies. Has she forgotten whose money that is? Consumers pay the carbon taxes passed through to them by industry. She's talking about spending consumers' money on subsidies for industry. Perhaps it's for a good purpose, though more likely it's largely gravy. But let's call it what it is—spending, gleeful spending.

The Danish subsidies are, she tells us, all for nice energy-related things such as "clean-burning fuel … and other environmental innovations." But whether the government buys clean-burning fuel or gives money to corporation X and says, "Buy clean-burning fuel," the government is spending the money. If the government itself fixes a road, that's spending. And if the government pays to have the road fixed, that's still spending. Governments almost always spend money on things that are good for someone. But that's not the same as saying, "Here's your money back; do what you want with it." That's what I call "tax and don't spend."

Prasad recommends a carbon tax, but she actually doesn't like it. She is not alone. The following is a common view among noneconomists:

> ▶ A carbon tax is expensive and won't do much.
> ▶ So spend the revenues on energy subsidies. Subsidies will work!

This policy has completely lost its way. Economists recommend a carbon tax because tax incentives work in a cost-effective way—much more so than subsidies. Economists know that just collecting the tax is not a social cost, provided the money is transferred back to consumers. Because subsidies are largely a waste of money, economists suggest either using the revenues to reduce some other tax or giving the revenues back directly. Economists definitely don't say to blow all the revenue on subsidies and pork.

Am I exaggerating? Does Prasad really think the carbon tax she recommends won't work? Her sole comment on the direct impact of a carbon tax is to say, "Higher gas taxes would raise revenue but do little to curb pollution." No, I'm not exaggerating. She reserves her praise for the subsidies and says nothing good about tax incentives. In her view the tax is just to "raise revenue" for the subsidy.

So why is collecting the tax so cost-effective? Since the tax works even with a 100 percent refund, it might actually seem to be infinitely cost-effective. The desired effect—energy conservation—appears to be completely free. This view has a grain of truth in it; returning the revenues does make the tax very cheap and cost-effective. But the tax has a cost that we need to understand, and I'll come back to that shortly, though I discussed it in detail in Chapter 16.

Here's the real reason the tax is so cost-effective. It works by getting consumers—you and me—and businesses to figure out how to reduce carbon emissions. You and I pinch pennies when it comes to energy spending, and so does business. But government subsidizers, under the influence of industry lobbyists, are spendthrifts. Even when the lobbyists are environmentalists, government officials can be spendthrifts.

Another view of the great cost confusion comes from the quaint early days of economics, when economists spoke of the "veil of money." Behind this veil lies the "real" economy, consisting of physical products and the actions of daily life. The great cost confusion comes from not seeing past the veil of money to the physical world. Look through the veil, and the carbon-tax revenues and refunds are seen as money changing hands and not as a net cost to the country. But the tax itself—regardless of what the government ultimately does with the revenues—has a real effect on the real economy, causing people to drive different cars, buy different light bulbs, and so on. The real costs of the carbon tax come from the real changes people make to save carbon. If people don't save any carbon, there are no real costs. When people do save carbon, they spend as little as possible.

The virtue of a carbon tax is that we all get to pick the best ways to save carbon, and we pinch pennies. So the tax incentives work cheaply. But are they strong enough? The incentive can be as strong as we want. For a powerful effect, we just set a high tax rate. With a 100 percent refund, the only social cost is the real cost of saving carbon, and there is simply no way around that cost.

The mistake of those who say, as Prasad does, that a carbon tax will "do little to curb pollution" is to see collecting revenue as a social cost. To people with that mind-set, even a weak tax looks expensive. And if we let them spend the revenues on subsides, the tax will be expensive. But if we avoid subsidies and turn up the tax while refunding 100 percent, the carbon tax will be the strongest, cheapest policy we can adopt. In fact, it is the only policy that is cheap enough to do the job without provoking a political backlash.

Can the Core Energy Plan Succeed without Subsidies?

Having seen that the government should avoid meddling, the question remains, how much should the government do? Is the untax enough? It fixes the underpricing of carbon and oil, but aren't there other problems with energy markets?

Chapter 7 lists four energy market problems that the policies of the Core National Energy Plan address. The underpricing of carbon and oil is first. But the second problem—OPEC's market power—is closely related and can be solved simply by adjusting the untax on oil to compensate for the distortions of the world oil market. No subsidies are needed.

That leaves two market failures that require a bit of government meddling—consumer myopia and a deficiency in advanced research. Chapter 20 discusses how to minimize meddling in the auto market while compensating for consumer nearsightedness. Chapter 21 discusses how to target government funding for advanced research and avoid unnecessary subsidies.

But can an energy plan with minimal subsidies get the job done? Even though subsidies are prone to corruption and expensive, aren't they the bitter pill we must swallow to bring about the enormous change in energy sources that we now require?

Since the untax and the race for fuel economy provide only broad financial incentives, many people think they won't work. They think financial incentives are soft and flabby and will be ignored.

Everyone could have ignored the oil in Pennsylvania, Texas, California, and Saudi Arabia. There were no government mandates to pump oil—only broad financial incentives. In fact, the whole world economy runs on financial incentives. So it's a bit ridiculous to call financial incentives weak. But perhaps they work for everything except energy problems.

OPEC's great energy experiment, which I describe in Chapter 8, proved that carbon pricing is powerful. OPEC's pricing reduced cumulative energy use in the United States by more than a decade's worth of present-day oil consumption. No other policy has come close to having such an impact.

In fact, it is subsidies that are weak and flabby because they are misdirected and often subverted. Prices guide the productivity and growth of the entire economy, and they can certainly change the direction of the energy market. The Core National Energy Plan avoids the great cost confusion and harnesses the power of a market economy to do our bidding.

~

The great cost confusion mistakes tax revenues for social costs. But fully refund those revenues, and they cancel, on average, the costs of the tax. And with equal-per-person refunds, the carbon tax still works its magic to reduce oil use and carbon emissions.

So take one step past the worst of the confusion, and we find a powerful policy with no net cost at all. This can't be right, but it brings us closer to the truth than believing it doesn't matter whether the revenues are returned to consumers or wasted on subsidies.

The truth lies beyond the "veil of money" and in the real world of choosing and using products and energy. Buying a hybrid car costs extra but saves on gas. The difference is part of the true net cost of an untax. The recycled tax dollars determine fairness but have nothing to do with the social cost.

See through the cost confusion, and we will adopt policies like those of the Core National Energy Plan, which are cheap and powerful. But with our vision clouded, we will see the tax revenue as an unavoidable cost and spend it all on subsidies rather than refund it. Our worst fears will come true as tax revenues that should have been refunded turn into the very costs we fear.

Part 4

Global Policy

chapter 23

Kyoto: What Went Wrong?

Clearly, more work is needed [on the Kyoto Protocol]. In particular we will continue to press for meaningful participation by key developing nations.

—Al Gore, *New York Times,* 1997

NINETY-FIVE U.S. SENATORS rejected a Kyoto type of treaty in July 1997, five months before 150 nations completed the text of the Kyoto Protocol—the actual rules for curbing emissions. The senators said they would not sign a treaty based on the protocol unless it imposed commitments on developing countries. They took a reasonable position, but one that closed the lid on a box the United States had built around the Kyoto process. No one conspired to build this box; it was just the result of unintended consequences.

Ironically, a great environmental victory in the early 1990s was the first step in constructing the box. As I discuss in Chapter 15, environmentalists and then-President George Bush ended a multiyear stalemate over acid rain by getting coal-fired power plants to accept emission caps imposed under a cap-and-trade policy. That success earned cap and trade the title of most successful market-oriented approach to emissions control. So when the U.S. team went to Kyoto, that was its proposal—to cap and trade greenhouse gas emissions. In the abstract, it made a lot of sense. But the countries of the world proved to be more complicated than coal-fired power plants.

Countries vary enormously in their levels of greenhouse gas emissions, so it's impossible to cap them all at the same level, and no one suggested that.

Instead, the treaty gave every country its own cap. That caused a lot of squabbling and naturally enough led to no caps for countries with low levels of per capita emissions—the poor countries. In effect, China, India, Brazil, and others argued that just because the rich countries started polluting first, they should not get to emit ten times more than poor countries, which have done less damage.

They have a point. But this leaves the Kyoto Protocol with an impossible contradiction. It's unfair to give poor countries caps that are five, ten, or even twenty times lower, on a per-person basis, than those of rich countries. But without such caps, poor countries have no obligation at all, and unfortunately, developing countries have the fastest-growing levels of emissions. China by itself emits more carbon dioxide than any other country, although its per-person emissions are low. Cap and trade sets up a clash between fairness and effectiveness. What is fair doesn't work, and what works is not fair. This is the box that the United States has built around the Kyoto Protocol.

This part of the book explains how to break out of the cap-and-trade box safely and effectively. In this chapter, I explain why we must abandon cap and trade as a global system before the world can solve the problems of climate change and energy security.

Not Fair

Caps on emissions are a burden, and the tighter the cap, the bigger the burden. On the other hand, getting a high cap can be worth a lot of money. That's because each country issues carbon permits up to its cap and can sell extra permits to companies in other countries for hard cash. In Europe people call this "selling hot air," and some Eastern European countries, including Russia, have lots of it to sell.

Russia gained a lot of its hot air by holding out and not signing the treaty until the country received an extra helping of free permits—that is, a higher cap. Because the United States would not sign the treaty, it could not go into effect without Russia's signature, which gave Russia a lot of leverage. This was a double win for Russia—the extra permits are valuable and they loosen the overall cap. As the world's number-two oil producer, Russia will be hurt by tight caps, which inevitably reduce world oil use and the price of oil.

The architects of the Kyoto Protocol may have issued permits unfairly, but this does not mean caps can't be fixed. Let's check to see if there's a way to patch things up.

The Kyoto Protocol sets emission caps relative to a country's emissions in 1990. In that year, the Chinese were emitting about 2.5 tons of carbon dioxide per person per year, and Americans were emitting about 23.4 tons per person. Even in 2008, India emits only 1.1 tons per person. I'm not criticizing Americans

Not the "You First Principle"

The *International Herald Tribune* called it the "You First principle" in June 2008 but says it's been the main reason for the Kyoto deadlock from the start. Developing countries say "You first," and we reply, "No, you first."

I've heard that's why the United States should pass one of the cap-and-trade bills before Congress. The problem seems easy to solve. We go first, and a year later they go. If they don't go, we have time to back out. Could such a little problem really explain a fifteen-year deadlock?

You would think that if You First is the problem, someone would say, "If you go first, then I'll follow." No one is saying that, particularly not the developing countries. They're saying, "You go, and we won't go." That's a problem that could cause a fifteen-year deadlock—or a fifty-year deadlock.

In June 2007, China published its *National Climate Change Programme*, which says China will follow the U.N.'s principle of "common but differentiated responsibilities."

China spelled out "differentiated responsibilities" for the Bali Climate Change Conference in December 2007: "The developed countries, whose emissions of GHGs [greenhouse gases] are the main cause of climate change, should have the primary responsibility to cut their high GHG emissions and to channel adequate financial resources and to transfer low-carbon technologies to developing countries. ... On the other hand, the developing countries, who are innocent in terms of responsibility for causing the problem, are by far the biggest victims."

That's it for developing-country responsibility. We "have the primary responsibility," and they "are innocent in terms of responsibility." That certainly is differentiated. But in case it's still not quite clear, Ma Kai, head of China's powerful economic-planning agency, explained it to the *New York Times*: "Our general stance is that China will not commit to any quantified emissions reduction targets."

This does not mean developing countries will not assume responsibilities. It just means they will not accept emission caps. Caps are out. They've been telling us for fifteen years, and they have good reasons. They are not going to change their minds just because we cap ourselves.

But I cut Ma Kai off in midsentence. He went on to say, "But that does not mean we will not assume responsibilities in responding to climate change." In fact, China is probably doing more than the United States is doing. It has adopted stricter fuel-efficiency standards, a more aggressive reforestation program, and a tougher energy-intensity reduction goal than George W. Bush's.

But according to that same *Tribune* article in June 2008, developing countries still won't "accept binding national emission caps." The trouble is not the You First principle. The trouble is caps.*

or complimenting the Chinese. This is just the situation we're in. It happens to make it impossible to set caps fairly and effectively.

To be effective, caps must be set low, near 1990 emission levels. That is possible, but how would such caps be adjusted? If India's and China's caps adjust down, they will forever be stuck a century or so behind us on carbon emissions. That's unfair. So their caps must be able to adjust up. But no one has figured out how to do that fairly and effectively.

Perhaps some mathematical trick would help us set fair caps. China could be capped relative to what it "would have emitted" if it had not been capped. The cap could be set farther below that would-have-emitted level each year—2 percent less, then 4 percent, then 6 percent, and so on. Unfortunately, as time goes on, we know less and less about what would have happened if China's emissions had not been capped. China might argue that, without a cap, it would have been emitting 127 percent more in 2010 than in 2000, as the DOE predicts. But environmentalists might argue that China would have been emitting only 27 percent more. That was China's emissions increase between 1990 and 2000. So China and the environmentalists might disagree by 100 percent on how tight a cap should be. Who is to decide?

If the cap is set 50 percent too high, it will have no effect. If it's set 50 percent too low and enforced, it will curb China's growth drastically. The latter outcome is unfair, and the former is ineffective. Caps based on predictions break down quickly. Besides, developing countries have been rejecting caps consistently and vigorously for fifteen years.

Billions Wasted

A working paper from two senior Stanford University academics, David G. Victor and Michael W. Wara, examined the U.N.'s Clean Development Mechanism (CDM) market for CERs (Certified Emission Reductions), which are international carbon credits issued by the U.N. The quantity of new CERs tripled in 2007 to a value of 12 billion euros.

They found that "much of the current CDM market does not reflect actual reductions in emissions, and that trend is poised to get worse." Moreover, investors paid roughly 4.7 euros for Chinese CERs corresponding to emission abatements that cost fifty times less.

Victor said, "It looks like between one and two thirds of all the total CDM offsets do not represent actual emission cuts," according to the *Guardian*, May 26, 2008.*

Caps Here and No Caps There

Without any way to cap developing countries fairly, the Kyoto Protocol takes another approach. It allows them to sell certified emission reductions, or CERs, to companies in capped countries. As the name implies, this is a certification that the uncapped country is making emission reductions that it would not otherwise make. I will call these CERs carbon credits.[1] A business in a country with a cap can buy carbon credits to help meet its permit requirement under its national

---

1. Similar certificates in other schemes are often called offsets.

cap-and-trade system. Each credit, like each permit, allows the emission of a ton of carbon dioxide or an equivalent amount of other greenhouse gases.

Reducing emissions in China is cheaper than reducing them in Germany, for example, so carbon credits save money and seem to be an excellent idea. And sometimes they work. However, no matter how well intentioned, credits will eventually run into two serious problems. First, they will cost a lot, and second, they will be gamed or cheated on.

Paying Others Is Expensive. To see how buying foreign carbon credits gets expensive, consider how things are going. In twenty-five years China will be emitting twice as much as the United States, Europe, and Japan combined. So if we do our part to buy China back down to our level, we will have to buy credits from China equal in amount to our own emissions. At $30 a ton, that would cost about $200 billion. I can't see us sending China that much money every year. That's more than $2,500 paid by a family of four.

Gaming with Carbon Credits. Gaming poses an equally intractable problem. And there is no way around it—it's just in the topsy-turvy nature of paying people *not* to do bad things.

For example, the operators of a coal-fired power plant in South Africa said they would keep using dirty coal unless they got carbon credits to buy some natural gas instead. But then someone found out that they had signed a gas contract before the CER policy went into effect. That is, they had already planned to cut their carbon dioxide emissions. They were simply hoping to defraud the United Nations, which administers the CER program.

Though someone detected the fraud in this case, eventually it will become impossible to know what the company would have done, because, with a carbon credit policy now already in place, the firm's operators have time to cover their tracks. If they plan to buy natural gas, they won't tell anyone until they lock in the credits.

This is why few markets sell negatives. People do plenty of annoying things, but rarely do we pay them $20 not to do this or $50 not to do that. Blackmail and protection rackets are two unpleasant exceptions.

In the long run, markets for not doing things just naturally end up in disarray. Say the city paid people for not parking too long in downtown parking spaces. You pull up to the curb, and the meter maid says, "If you leave in less than an hour, I'll give you $2." So you do, and she does. But when you get home, you tell your teenager about this, and the wheels start turning. Pretty soon your kid parks downtown, leaves his parking space after ten minutes, and collects $2. He then parks two blocks away and collects $2 more, and so on. Pretty soon downtown has turned into a game of musical cars for teenagers. The payments are for leaving parking spaces, but the result is parking spaces mobbed by teenagers.

Perhaps you still think people wouldn't do things like that or that we could catch them. But consider this example: Certain chemical plants around the world emit just about the worst greenhouse gas imaginable. The refrigerant HFC-23 is 11,700 times worse than carbon dioxide. But a European company can pay a chemical plant in China to stop emitting HFC-23. The Chinese plant puts the gas through an incinerator to avoid emitting it into the atmosphere. Incineration is a cheap process, and for every ton a plant burns it earns 11,700 tons of carbon credits, which the European company purchases. In early 2008, international carbon credits were worth about $25 per ton. So incinerating a ton of HFC-23 was worth close to $300,000, while incineration cost only about $5,000. Most of the credits granted in the first few years of the CER program have been for HFC-23 incineration.*

So how is this story like the one about the teenagers parking downtown so the city can pay them not to? There are rumors that Chinese companies have built chemical plants mainly to cash in on carbon credits.

But even if no one intends to misbehave, the CERs encourage it. Whoever takes most advantage of them makes the most profit and can sell their product for less and undercut their competition. Businessmen fear their competitor will employ such a strategy, and so, in self-defense, they feel they must employ it themselves. Paying for negatives—giving out carbon credits for not emitting—can corrupt honest people.

In fact, the United Nations has known of the CER problem from the beginning and terms it "additionality." That is, the United Nations requires projects to be "additional" reductions to emissions. Now my copy editor asks "additional to what," and that is exactly the question the United Nations did not, and can never, answer clearly. The answer will always be, "additional to some hypothetical future world." The idea of enforcing an "additionality" requirement is just wishful thinking.

Just for comparison, consider what would happen if instead of the United Nations giving China carbon credits, China had agreed to put a tiny $1-per-ton tax on greenhouse gas emissions. That would mean $1 per ton of carbon dioxide and $11,700 per ton of HFC-23 emissions. That's more than it costs to incinerate HFC-23, so chemical plants would incinerate and pay no tax at all. In fact, many developing countries—and, to some extent, the United States as well—subsidize fossil fuel. A requirement to stop subsidizing greenhouse gas emissions and to impose even a small tax would be a huge step in the right direction—not least because developed countries would then have to meet their caps by cutting emissions at home.

Charging people who park too long is a better idea than paying them to leave sooner. Every city in the world has figured this out. The same principle holds for taxing emissions instead of paying people not to emit. Sooner or later, this will become all too apparent.

Avoid Global Emission Caps—Require Equal Effort

Capping emissions country by country boxes us in. It's unfair to cap poor, rapidly growing countries. But paying them not to emit is too expensive for the rich countries because of waste and overpayment. We need a fair and effective way to include the developing nations. Since caps don't work, the obvious alternative is carbon pricing. In fact, that should have been the first choice.

Instead of a requirement that every country stay under a certain cap, the rule would be that every country must put a certain price on carbon. Countries could achieve that price with a cap, a tax, or an untax. Each country would be free to choose. Global carbon pricing is inherently more fair because it requires a level of effort instead of a specific cut in emissions.

If your family is weeding the garden, a requirement that each person pull 30 pounds of weeds may be next to impossible for the little kids. But a requirement that everyone pull weeds for thirty minutes may be reasonable. In any case, it has a better chance of being fair.

A carbon price of $30 per ton scales automatically to a country's carbon level. In a country where people use 1 ton per person per year, the average cost will be $30 per person per year. In a country where people use 20 tons, the cost will be $600 per person per year. Of course, the money stays in the country, so this is not a cost to the country. The government can, if it wishes, give it all back—via an untax or another method—as long as it does not reward those who emit more carbon. If a nation adopts an untax, it helps the poorest people in that country.

This approach ensures that carbon control does not limit economic growth. With a carbon price of $30 per ton, nothing stops India from becoming richer than the United States. But if India's emissions are capped at their present level, it makes it almost impossible for India to catch up economically.

At this point, some people will conclude that carbon pricing seems more fair simply because it's weaker. But that is not the case. As I explain in Chapter 18, a cap that causes a $30 carbon price has exactly the same effect as a $30-a-ton carbon tax. Both a cap and a tax put a price on carbon, and the price—and nothing else—does the work. A cap is only stronger if it tricks the world into accepting a higher carbon price. But the opposite is more likely. People are afraid a cap might push carbon permit prices too high, so they set caps cautiously and build in loopholes. In any case, if caps push carbon prices to $100 while the world is only willing to accept $50 carbon prices, the world will change the cap and not the other way around.

~

It would make little sense to suggest such a radical new course—global carbon pricing—if the old system of national carbon caps were viable or needed only minor adjustment. But an international system of capping is not an option. That does not mean individual nations need to stop using cap-and-trade systems. Nations can still choose any method they want to raise their national carbon price to the global-carbon-pricing target.

For good reasons, developing countries will not accept internationally set caps. Paying them to curb emissions will prove too expensive, especially because payments not to emit are ineffective and inevitably lead to gaming and fraud. Fortunately, a global carbon price can provide a fair and effective standard, and it is the best hope for international cooperation.

chapter 24

Global Carbon Pricing

We have everything we need to get started, save perhaps political will, but political will is a renewable resource.

—Al Gore, Nobel lecture, 2007

HALF THE WORLD will not accept carbon caps but might accept a carbon price requirement. Such a requirement would not put a lid on growth in developing countries. And, if the requirement was too burdensome on poor countries, they could be compensated by international payments. Individual countries could choose caps, taxes, or untaxes at the national level.

Countries that are particularly dependent on oil would be free to target carbon from oil. Targeting oil would decrease political resistance and increase the policy's effectiveness at reducing oil prices. As Al Gore says, the world may not yet have the political will to get started. But that could change if people begin to see the benefit of cooler global temperatures combined with the benefit of lower oil prices. Political will is most effectively renewed with a dollop of financial self-interest.

Switching from Kyoto's caps to a new, global-carbon-pricing policy will require a major reorientation of the Kyoto Protocol. In this chapter, I describe a basic design that, because of its flexibility, requires only minor adjustments to existing national carbon control policies. I present a simplified version of the design in this chapter, adding modifications for fairness and enforceability in Chapters 26 and 27.

The Price of Carbon

Global-carbon-pricing policy sets a target global carbon price and then makes sure the world achieves it on average. To make the policy flexible at the national level, the global carbon price must be defined to work with any type of national carbon policy—cap and trade, gas tax, untax, or any other method of making carbon expensive. To achieve this flexibility, global-carbon-pricing policy defines the national carbon price as the average carbon price over all fossil fuels and does not apply the requirement to every individual purchase of fossil fuel.

Price is just revenue divided by quantity sold. Collect $100 from selling ten items, and we know the average price is $10 per item. The national average price of carbon is total annual revenues from carbon charges divided by total carbon emissions during a year. So if the United States collects $60 billion in carbon charges in a year and emits 6 billion tons of greenhouse gases, our national price of carbon is $10 a ton. That's all there is to it.

Well, not quite. Suppose a nation's carbon cap-and-trade program gives away all its permits to coal-fired power plants—not a good idea, but just suppose. How should a global-carbon-pricing policy give that country's program credit for carbon pricing? The global policy must work with any national carbon policy to avoid giving any country an excuse to opt out.

Because free permits given to coal plants collect no carbon charges for the government, it seems as if they should not contribute to the national carbon price. But if permits given out for free cost $20 a ton in the private market, they put just as much pressure on companies that need them as a $20 carbon tax. So these permits should get just as much carbon pricing credit. This is fair and easy to arrange. Carbon permits receive carbon pricing credit equal to their value at the time they are retired to cover emissions. If a million permits are retired in May and the average price in May is $30 per permit, the country receives credit for $30 million of carbon pricing revenues.

Carbon taxes, gas taxes, and untaxes all collect revenues that are easy to count. Subsidies for ethanol and wind will be unnecessary once fossil fuel costs more. However, if countries still offer such subsidies, they should not be counted, because the track record of subsidies around the world, including in the United States, is dismal. In fact, an enormous benefit of global carbon pricing is that it dramatically shrinks wasteful energy programs.

Carbon or Greenhouse Gas?

Fossil fuel accounts for about 70 percent of greenhouse gas emissions. However, we should not ignore the other 30 percent. Carbon dioxide is the greenhouse gas emitted when people burn fossil fuel. Since this book is about energy policy, I'm most concerned with carbon.

But sometimes people use carbon to refer to all greenhouse gases. For example, Europe's greenhouse gas markets are called carbon markets. In that tradition, when I speak of carbon, in most cases I mean all greenhouse gases.

Appliance and fuel standards should continue, but to be counted they would need to be converted to a feebate, or efficiency race, approach.[1] This will improve their performance and reduce their vulnerability to bureaucratic foot-dragging. In Chapter 20, I explain how to do this for fuel-economy standards, and the same techniques work with appliance standards.

Each country would have to count total emissions and total revenues to determine its national carbon price. And greenhouse gases come in many types and from many sources. In the United States, the Environmental Protection Agency and the Department of Energy keep track of these, but would this be possible in most other countries? What about Estonia, Slovenia, and Romania? As it happens, by May 2008, these and thirty-six other countries had already filed their national greenhouse gas inventories with the United Nations under the Kyoto Protocol. This information is necessary under any climate-change protocol. So the difficulties with measuring emissions appear surmountable and, in any case, cannot possibly be an argument against switching to a global-carbon-pricing policy.

In fact, permit prices under cap and trade are more sensitive than is carbon pricing to errors made in counting emissions. The European Union miscounted by a few percent during the trial carbon-cap period before 2008 and issued a few too many permits. The result was that the price of carbon crashed from about $30 a ton to under $1 a ton. Under a global-carbon-pricing policy, such a small mistake would cause only a small problem. A similar-size error might cause a country's carbon price to be miscounted as $30 per ton instead of $29.

Who's for Global Carbon Pricing?

Global carbon pricing goes by several names and has probably been outlined a hundred times. Several of its advocates stand out because they have written about it as a solution to the Kyoto difficulties.

These include William Nordhaus of Yale, Richard N. Cooper of Harvard, Nobel Prize winner Joseph E. Stiglitz, and the recent chairman of the president's Council of Economic Advisers, N. Gregory Mankiw.*

I describe a specific approach to global carbon pricing in this book, but whenever I say someone else favors global carbon pricing, I only mean they favor the generic concept, not my specific proposal.

The World Bank tracks most countries' finances closely. So keeping track of carbon price revenues should not be difficult. Moreover, low-emission countries will want the fairness payments discussed in Chapter 27, so they have reason to cooperate.

To sum things up, a nation's carbon price is its total carbon pricing revenue divided by its total greenhouse gas emissions. Revenues are not hard to count, and countries must count their carbon emissions under any system that works.

---

1. As I explain in Chapter 27, fairness payments, a part of the proposed global-carbon-pricing policy, will indirectly reward successful appliance standards even if they do not contribute to carbon revenues and so are not counted toward global carbon prices..

Flexibility and Fairness

Global carbon pricing does not require that all carbon be priced the same. The price can vary from one type of fuel to another. That provides flexibility in the design of national policies. Similar to the international payments for carbon credits under the Kyoto Protocol, "fairness payments" can be arranged to compensate poor countries for full participation in global carbon pricing.

Flexibility. Countries can, if they like, tax oil carbon at a high rate and other carbon at a low rate, just as long as they collect enough total revenue. Because taxing oil reduces its use, it also helps to lower the world price of oil. Large oil-consuming nations can have enough of an impact to save a significant amount of money with this approach. Europe already does this, and it makes sense for the United States as well.

But when one country uses less oil, all consuming countries benefit, which argues for cooperation. A global-carbon-pricing policy makes it easier to price oil high and thus encourages cooperation among oil-consuming nations. Since a global-carbon-pricing policy requires countries to collect a certain amount of revenue from pricing carbon, why not focus much of the revenue collection on oil? That has the fringe benefit of reducing oil prices.

In effect, flexible global carbon pricing encourages the formation of a large international consumers' cartel. As I will discuss in more detail later, this is not just good for energy security. The benefits of such a cartel would also provide much of the glue that will hold together an international climate agreement.

Fairness. Although a country with one-tenth the income of a rich country would pay about one-tenth the carbon charges if its carbon prices were the same, this is probably not fair. Generally, poor people find it harder to give up a certain fraction of their income than richer people do. Moreover, they have caused much less of the problem, whether we consider climate or energy security.

Fair carbon capping requires taking into account a host of considerations, and even then we end up with a stalemate. Fair carbon pricing is simpler and is best achieved with fairness payments. A fair formula can be based on just one factor, such as per capita income or emissions per capita. This avoids all the bickering over individual caps. In Chapter 27, I provide a specific fairness design that also provides a second reward for reducing emissions.

In any case, the most important point regarding fairness is that it is easy to address within the framework of a global-carbon-pricing policy. Global carbon pricing automatically takes care of the overwhelming fairness concerns that block the road to carbon caps. Global carbon pricing also provides the flexibility to address the simpler fairness concerns that remain.

Even Pricing at Zero Would Be a Step Forward

As I mentioned earlier in this chapter, one of the greatest benefits of a carbon pricing policy is that it reduces wasteful energy programs around the world. And the most wasteful of all such programs are fossil-fuel subsidies. These cost governments more than the benefits they provide, besides hastening global warming and decreasing energy security. Subsidizing ordinary goods wastes money, because it causes people to overuse the subsidized goods. Subsidize wool, and people wear wool instead of cotton, even though wool costs more to produce and even in cases where cotton works just as well.

When countries subsidize fossil fuel, they waste money, damage the climate, and increase energy insecurity. That's a lose-lose-lose policy. Global carbon pricing puts a stop to such policies worldwide, saving the world hundreds of billions of dollars a year. With a carbon pricing requirement, subsidies count as negative pricing. So to achieve the required price, a country must abandon carbon subsidies or apply an extra-heavy carbon tax that counteracts the subsidy.

Fossil subsidies also cause problems on a global scale. In mid-June 2008, China raised its domestic price of gasoline, and the *New York Times* reported that: "The price of light crude fell $4.02 to $132.66 a barrel following [China's 16 percent] fuel price increase announcement. … After the hikes, prices [in China] rose to about $3 a gallon. … In 2007, China's subsidy of gasoline alone was $22 billion, close to 1 percent of its gross national product." If reducing the subsidy cut world oil prices, then the subsidy itself has been raising them.

This little report speaks volumes about Kyoto and the need for a global-carbon-pricing policy. Under Kyoto, China is spending a good fraction of what an effective anti–global-warming program would cost on subsidies that exacerbate global warming. Along with some positive programs, China is doing the exact opposite of what the subtitle of this book recommends. China's subsidy policy can be summed up as "how to wreck the climate and help OPEC charge the world more."

The market's reaction to China's price hike gives us some idea of the past cost of its subsidy policy, but only a hint, because China's price hike will take years to fully reduce Chinese oil consumption. China cut its gas subsidy by about forty-five cents a gallon, and that immediately saved the world $4 per barrel. That comes to about $100 million a day saved on OPEC's exports, which is $36.5 billion a year. And, of course, you can triple that if you want to add in all the other oil companies—including Exxon and the Russian companies.

China's domestic oil price increase does not likely indicate a phaseout of gasoline subsidies. Rather, it's an indication that the country's subsidies had gotten out of hand. The Kyoto Protocol has handed China hugely profitable carbon credits. At the same time, the protocol allows the Chinese to turn around and increase their damage to the climate by subsidizing oil imports—imports that increased 25 percent in the year ending May 2008. A global-carbon-pricing policy, on the other hand, requires all countries—including the United States, China, Saudi Arabia, and Iran—to stop subsidizing fossil fuels and start taxing them.

~

Global carbon pricing allows flexibility in the design of national policies. The only requirement is that the combined policies of a country collect enough revenues from carbon pricing to meet the global pricing target.

This flexibility encourages countries that import a lot of oil to cooperate in setting a high price on oil carbon, which helps reduce world oil prices. This makes global carbon pricing the best policy for energy security as well as climate stability. The synergy between these goals provides a strong incentive for international cooperation, as I discuss in Chapter 29.

chapter 25

Does the World Need a Cap?

The Intergovernmental Panel on Climate Change says we must reduce carbon emissions 80 percent by 2050.

—Environmental urban legend[1]

ENVIRONMENTALISTS OF A CERTAIN STRIPE are saying there's a scientific consensus that we must reduce carbon emissions 80 percent by 2050. But the Intergovernmental Panel on Climate Change (IPCC)—the U.N.'s climate science group—has said nothing of the kind.* The IPCC does predict the global, but not national, emission levels that would hold greenhouse gas concentrations down to 450, 550, or 650 parts per million (ppm). But they haven't said which target we must shoot for. The current carbon dioxide level has already reached 380 parts per million from a historic starting level of 280.

The legend contains a nugget of truth, reflecting a popular environmentalist choice of 450 parts per million as a target. Some reports, which the IPCC has summarized but not endorsed, say that the developed countries must push their emission levels down to 80 percent below 1990 levels by 2050 to make up for what the rest of the world is likely to do—if we want to stabilize greenhouse gas concentrations at 450 parts per million.

---

1. Because the author of this quote has recanted, I will not disclose his or her identity. But a large number of people still believe the legend.

While the IPCC does not recommend any particular level of GHG concentration, it does tell us something about what the levels mean. In particular the 450 ppm target corresponds to an equilibrium temperature increase of 2.1 degrees centigrade. What goes unmentioned by those advocating this target is that the this equilibrium increase will only be attained after about 500 years. If we were satisfied with what the IPCC calls scenario B1, which corresponds to a temperature increase of 2.3 degrees centigrade in about 2100, a global emissions increase of 40 percent in 2050 above 2000 levels would be feasible (see endnotes).

Yes, that's a 40 percent global increase in emissions under the B1 scenario, and an 80 percent decrease for developed countries under the 450-ppm equilibrium scenario. Under the first, the temperature increases 2.3 degrees by 2100 and under the second, 2.1 degrees eventually. These are a bit difficult to compare because, even if the world increases emissions by 40 percent, the developed countries might need to reduce emissions to compensate for increases in the developing countries.

Now, I'm not saying the 80-by-2050 target isn't fair or that it's not a good idea. Perhaps it is. But it is wrong to say there is a scientific consensus for the very-long-run 450 ppm target. It is popular with quite a few scientists, but this popularity is based on value judgments as well as on science. The IPCC itself simply lists this target as the most stringent one studied among all targets studied in the 177 reports it reviewed.

Chapter 23 concluded that national caps are out of the question as a comprehensive method of global organization. So, as long as so many people in developed countries feel caps are the only means to achieve success, we will probably make little progress toward a workable solution. In this chapter, I argue that internationally-set national caps are not necessary and, in fact, do not provide the kind of certainty that people hope for. Since they are out of reach, this is not bad news. The good news is that carbon pricing would work about as well as caps are imagined to work, if we did agree on where we're going.

When Is a Cap Not a Cap?

Cap and trade is supposed to work by setting one big cap for all emitters combined. With a national cap-and-trade program, individual companies don't have caps. With a global cap-and-trade program, you'd think individual nations would not have caps. You'd think the whole world would just have one big cap.

But we don't have a world cap. Instead, we have lots of caps for individual countries. What's going on? Under the Kyoto Protocol, it's a bit mysterious, with some countries capped and others not. But suppose every country had a cap. Would that make a world cap? It would, and the sum of all the country caps would be the world cap. But what is the effect of the "national caps"? Do they do more than just add up to determine the total? If that were all they did, countries would not worry about their particular cap, but here's why they worry.

"Capped" countries are allowed to trade carbon permits. This saves money because countries for whom it's expensive to reduce emissions pay companies in other countries to reduce their emissions. That generally saves money, and it's a good thing. It's already happening, and more of it will happen in the future.

But international trading means the national caps are not actually caps at all. Any country can emit more than its cap just by purchasing some permits. It's exactly like a cap-and-trade system on U.S. companies. None of them get individual caps; any company can buy extra permits if it needs them.

When a company buys extra permits, it's because it does not have as many as it needs. When a nation buys extra permits, it's for the same reason.

What people call national "caps" actually function as allocations of free permits. But no one calls them that, because giving out free permits to emit greenhouse gases just doesn't sound right.

If Germany's cap is 1 billion tons, that means the Germans have permission to emit up to 1 billion tons without paying for extra permits. That's exactly the same as if Germany were given free permits to emit 1 billion tons. So a national cap does not cap a nation; it tells the nation how much it can emit for free before it has to start buying permits from other countries. In effect, national caps are just allocations of free permits. That's why countries care a lot about where their cap is set.

It's All about Free Permits

If the United States adopts a cap that slides down to 80 percent of 1990 emissions by 2050, we will almost surely join the world trade in carbon permits. Once our so-called cap becomes tough to meet, companies will apply enormous pressure to be allowed to buy cheap international permits or carbon credits, which are like permits but from countries without caps. Businesses will point out that this saves money and helps poor countries.

Once carbon trading starts, as I just explained, we will no longer have a cap of 80 percent in 2050. Instead, we will cut emissions by perhaps 40 or 50 percent and buy permits to cover the rest.

Massachusetts Institute of Technology (MIT) researchers came to similar conclusions when they modeled actual bills before Congress. In the MIT model, though, companies saved up permits in the early years and then used them to circumvent the cap in the later years. When businesses can bank permits, national caps don't come close to telling us how much a nation will emit in 2050. The same is probably true when we allow international permit trading.

So capping at 80 percent below our 1990 level means we give ourselves very few free permits and buy the rest from abroad. But we do not really cap ourselves at 80 percent of 1990 emissions.

Capping Half Doesn't Cap the Total

If all other countries had caps, they would have to emit less when we bought their permits from them. But under the Kyoto Protocol, most of the world's emissions will not be under caps by 2050. Here's what will happen when we buy permits from abroad.

We might buy them from Europe. Then European companies will be shorter of permits, and they are already short. So they will buy carbon credits from, say, China. China will emit less and sell the European companies their credits. But as we saw in Chapter 23, we have no idea what China will be emitting in 2050. Less than what it emitted in 1990? No. Less than it would have been emitting in 2050 without the Kyoto Protocol and its successor protocol? Supposedly. But we can't even make a good guess about that. As we conserve and lower the price of oil, China will use more oil. Perhaps if we had not paid them to use less coal, their own pollution would have driven them to it.

As a result, under a partial cap system—which is the best we can hope for—the world will not be capped. The United States will not be capped, because it can buy permits from abroad. And the net result is … who knows what?

If the United States does cut its emissions by 80 percent while the world as a whole increased its emissions by 40 percent—as would happen under a scenario like IPCC scenario B2—where will it get us? We will be down to about 2 percent of world emissions, and something like 90 percent of emissions will come from the developing countries, which made no commitments under the Kyoto Protocol.

Hitting the Target

Could we hit a target if we did cap the whole world? In theory, yes—a cap set equal to our target should assure we hit the bull's-eye. But before we check on that theory, what if we did hit our target? Would the target necessarily be the right one?

In fact, it would certainly be the wrong one, and probably by a lot. Scientists are pretty sure that humans are causing most of the global warming. But nature also causes warming and cooling, and it's hard to predict that, not to mention the impact of human emissions. In fact, even assuming we stabilize greenhouse gas concentrations at 450 parts per million in 2100, climate science gives us only a fifty-fifty chance that the globe will warm up by less than 2 degrees centigrade. It could turn out significantly warmer or cooler. For one thing, climate change depends on the role of clouds, and science will not understand their impact well for at least a decade. Then we might get good news or bad news.

Now, forget for a minute that we have a moving target and assume we know the ideal target for 2050 exactly. Will a cap set in the next, say, four years get us there? No. Even if I am wrong about China and India, and they will eventually accept a cap, they will want to start slowly just like the developed countries did. That means at least fifteen years of developing and testing caps before getting down to ones tough enough to do much good.

In contrast to China and India, both of which want to curb global warming, countries like Saudi Arabia, Iran, and Russia certainly do not want to see a successful climate policy. That would mean cutting world oil use—and their profits. Unfortunately, these countries use a disproportionate amount of the world's oil. We can be sure they will not accept an appropriate cap for decades.

So it's exceedingly unlikely that we'll ever see a cap on the whole world's emissions, and the capping process would be slow and laborious at best. We don't even know the right target. And if the Kyoto agreement is any indication, quite a few countries will fail to meet their caps.

All this doesn't prove that using caps is a bad idea. But it does appear that using caps will require many rounds of adjustment and that the adjustments will go on indefinitely.

Hitting a Target by Setting a Price

Could we use carbon prices to hit an emission target? Any policy we choose will be fraught with missteps and adjustments, learning and forgetting, cheating and technical breakthroughs. So forget the simplicity of picking the right number once and for all. It's not going to happen. We cannot know the future; we only know we are at risk. But that doesn't mean we can't achieve our goal.

Since we cannot know the future, we need a policy that adjusts easily and that doesn't frighten people with unknowable costs. A global carbon price comes closest to filling this bill.

Easy to Agree On. A global-carbon-pricing policy requires a single target price for carbon. Compromise is required to reach agreement, but the simplicity of global carbon pricing makes this easier. Also, carbon pricing is designed to offer the carrot of lower world oil prices as an inducement for nations to cooperate.

With individual caps, every country has an incentive to fight hard for a lenient cap. Remember that national caps are really free carbon permits, and carbon permits are worth a lot of money on the worldwide permit market. In hammering out the Kyoto Protocol, countries have constantly struggled over individual caps. Handing out trillions of dollars to 180 countries on the basis of negotiation, as opposed to a rule, is the best way I can imagine to cause endless bickering.

Simple to Adjust. When the carbon price needs adjustment—and it will—only one number needs changing. The fairness rule, described in Chapter 27, automatically adjusts with the global price level. So that one value is all that needs adjusting. The policy leaves no room to bicker for individual advantage.

When nations negotiate a global price, everyone is in the same boat. If the price goes up, it costs everyone more, but it also improves climate stability and energy security for everyone. Countries will still have disagreements, but no country can get the benefit of a stronger policy without contributing its share.

Doesn't Frighten People. Different people think in different ways. Environmentalists think the environment is a top priority. So they think first of emission limits and give little weight to the cost of meeting them. They are not frightened by the cost uncertainty of caps.

Most people, even rich Americans who tell pollsters the environment is important, are not yet willing to put much money on the table. When pollsters ask if they will accept higher gasoline prices, they say no, often rejecting increases that would cost just a few dollars per year. In poor countries, this is even more true.

I'm not concerned with who is right but about how to get the job done in the real world. The project of turning most of the world into dedicated environmentalists may eventually succeed but not in time to stop global warming. Instead, we must work with the situation at hand.

The points I make in Chapter 15 about concerns with the cost of caps carry even more weight when it comes to developing countries. For the vast majority of the world's citizens, immediate costs come first and long-term benefits second. These people are willing to spend something, but they do not want to lock into a cap with little idea of what it will cost them. Capping fossil-energy use at levels equivalent to those found in the United States in the late 1800s can be a frightening prospect. I personally think we can achieve such reductions at a surprisingly low cost. But it has never been done, and people are frightened to commit to such an experiment, especially when they are poor.

This is roughly how many people look at the problem:

> You want me to reduce my fossil-fuel use to the level Americans used in the nineteenth century? Some say that will be cheap, and others say we must sacrifice a great deal for the environment. But no one really knows what's necessary. I'm willing to start. But don't ask me to lock into a plan when I don't know either the cost or what is necessary for the environment.

The bottom line is that caps frighten most people once they take a close look at them, because no one knows what caps will cost. If people are forced to accept a cap with an unknown cost, they will fight for a weak cap. In the end, people will accept a stronger policy, even if it costs more, to avoid one that's cheaper but risky.

~

Although caps appear to be a certain means of attaining climate goals, locking in a goal regardless of cost makes the policy's cost highly uncertain. For most of the world's citizens, unknowable short-run costs are more troubling than uncertainty in the distant future.

So the world is not about to cap emissions, and without a worldwide cap, national caps are simply a way of dividing up costs. They shift income from tightly capped countries to countries with looser caps or no caps. This turns a cap into a punishment, and a lack of a cap into a reward, and reinforces the resistance to caps.

A global carbon price has the opposite effect. It puts all countries in the same boat. Raising the price of carbon costs all countries the same proportionally and increases climate stability and energy security. A country cannot use the system to help itself at the expense of other countries. All nations rise or fall together. This way lies cooperation.

chapter 26

International Enforcement

What you cannot enforce, do not command.

—Sophocles (496 B.C. to 406 B.C.)

WHY DO PEOPLE DRIVE ABOUT 70 MILES AN HOUR on the freeway? Because the speed limit is 65. Actually, that's not exactly why, and the little 5-mile-an-hour discrepancy gives us a clue. It's the police and the courts that keep most of us from speeding, not the limit itself. The police don't usually ticket you till you are driving about 10 miles an hour over the limit. That, and a bit of caution, explains the 5-mile-an-hour discrepancy.

You may be thinking this is pretty obvious, and it is. But people constantly forget about it in discussions of international policy. The authors of the Kyoto Protocol set speed limits—caps—but forgot about the police and the courts. This works to some degree with a small group of cooperative players, such as about half the nations of the European Union. But bring an outlaw nation such as Canada into the mix, and speed limits without police are a joke.

OK, Canada is hardly an outlaw nation, and that's my point. Canada is one of the most cooperative nations in the world, and a liberal, pro-Kyoto government was in power during the crucial period when nations were hammering out the protocol. But the country is still exceeding its Kyoto limit by something like 20 percent. Think what will happen once we include a lot of countries that are less cooperative and enthusiastic than Canada and when the requirements get tighter.

Keeping 180 nations in line requires an effective enforcement mechanism. Doing without one is completely irresponsible. But enforcement need not be heavy-handed. The penalties only need to be strong enough to compel an average level of compliance, because only average emissions and average oil consumption matter for global climate change and energy security. In this chapter, I show how to enforce a global carbon price effectively but with the lightest possible touch.

Before we discuss how to enforce a global "speed limit," though, we need a clear picture of exactly what a carbon speed limit looks like. The global carbon price determines the "speed limit" for each nation. If that price is $20, and a country emits 1 billion tons of carbon dioxide per year, its annual target revenue is $20 billion dollars. That's all that must be enforced on average for all nations—their target revenues.

Light but Effective

The first principle of gentle enforcement is that it's OK for a country not to achieve the target price. However, in that case, the country must pay a fine. In other words, countries can buy their way out. Some people will prefer a more moralistic approach, but as we saw in Chapter 17, this benefits no one and complicates the system. A carrot-and-stick approach of fines and rewards will make the system more popular with both those buying their way out and those getting rewards. And this flexibility will not hurt the outcome one bit.

The second principle of gentle enforcement requires that fines exactly pay for rewards. Revenue from fines should not be used to pay for other projects, because this will prove costly and cause resentment. This is, of course, the classic feebate mechanism—that ugly word again—which I have recast as a race. You also hear this approach called a revenue-neutral mechanism. It's a popular design because it works so well and so simply; it causes no fights over where the money comes from or who should get it.

Enforcement as a Race

As with the race to fuel economy, it helps to think of the enforcement rule as a race—in this case, a race to higher carbon prices. The winners earn rewards, and the losers pay for the prizes, so everyone is motivated. In this race, each country's score is its actual revenue collection minus its target revenue collection—that is, actual carbon revenues minus what the country would collect if it set its carbon price equal to the global target carbon price. Collecting too little revenue gives a country a negative score.

We also need a simple rule for handing out prizes. Remember that negative prizes—fines—pay for the real prizes. First, we set a prize rate, Z, which might be, say, 20 percent. Then, if a country collects an extra $500 in carbon revenues, its prize is $100. If collects $500 too little, it pays a fine of $100.

If Z is too high, countries will overcomply to earn the big prizes. And if Z is too low, many nations will undercomply because the penalty is too small to worry about. So we can attain average compliance—which is all we need—by choosing the right prize rate, Z.

That's a simple idea, but what if most countries do better than required or if most countries do worse? Then either the fines would not pay for the prizes or we would collect more fines than we need. Even with a good system for choosing Z, this can happen. Keeping revenue neutral requires adjusting the fines and the rewards when, at first, they don't balance each other out.

If we collect too much in fines we can just refund the extra proportionally to all countries. If we collect too little, it works a bit differently. We simply divide the fines among the winners in proportion to their scores.[1] In either case, the fines exactly pay for the prizes.

Adjusting the Prize Rate

The enforcement system I just described works fine provided the prize rate, Z, is strong enough but not too strong. Economists should be able to make a reasonable first guess at Z. After that, administrators will have to adjust the rate. However, a simple rule can determine how Z adjusts. If the weighted global average carbon price is only half as high as it should be, then the next year Z doubles to provide twice the incentive. If the average carbon price is 30 percent above target levels, then Z is reduced by 30 percent.

That's all it takes. Enforcement won't be perfect. Some years the carbon price will be a bit high, and some years it will be a bit low. But, on average, it will equal the carbon price target. This means that the global carbon price will be accurately enforced—on average. Global warming is a slow process, and there is no need to be right every year.

How Big a Fine?

Would a government collect $10 billion with a carbon tax to avoid $9 billion in fines? Wouldn't the fine need to be $10 billion—100 percent of the tax collected—to get reluctant countries to comply? Not at all. If a country collects $10 billion in revenue, it can refund all of it to its citizens while a $9 billion

---

1. This small change in the case when fines don't cover the prizes assures that any nation that sets exactly the global carbon price will never have to help pay for prizes. This is done to prevent any perception of unfairness.

fine leaves the country. So the fines do not need to be so high. Even a $1 billion fine or less may well encourage a government to collect and refund $10 billion with a carbon tax, untax, or a cap-and-trade system.

So a low prize-and-fine rate, even 10 percent or less of revenue collected, is likely to motivate compliance. This is good, because big fines are unpopular, even when countries deserve them. Of course, any country can avoid a fine simply by setting its carbon price to the global target level—or a little higher to get a reward.

Enforcing the Enforcement

The police enforce the speed limit by handing out tickets—little slips of paper that you can just tear up and throw out the window. So obviously the police are not enforcing the speed limit at all. Well … actually they play a crucial roll, but without backup their tickets would do no good. To get people to drive slowly, we need three layers: the speed limit, the police, and backup enforcement with real muscle—prison or wage garnishment.[2]

Few people get their wages garnished for speeding. But garnishing is still part of the system. It's the threat of garnishment that does the job, and because it's a credible threat the government almost never needs to follow through. But without some real threat, the whole enforcement system is a joke.

The same holds for an international climate agreement. Without a real threat, countries will miss their targets and be issued tickets and throw them out the window, so to speak. More likely they will say, "Oh, sure, we will pay," but they will never get around to it. After a while, some countries will see they don't have to pay the fines, and they'll start missing targets regularly, but not by too much.

But then the leaders of other countries will think, "If they are going to miss their target by 20 percent, we are going to miss ours by 20 percent." And eventually the whole system falls apart.

I cannot prove this will happen. But an organization with members who cooperate for mutual economic benefit is basically a cartel. And economists have studied cartels for a long time, and the main thing they've learned is that cartels tend to fall apart. And the bigger they are, the faster they fall apart. Once a cartel has more than a hundred members, it's likely to fall apart before it ever gets organized. The trouble is that standard cartels can make their own rules (speed limits), but the only real enforcement they have is this: If one member cheats, the others can shut the cartel down and punish everyone.

---

2.   Even the threat of taking away a license won't work, without backup enforcement for that penalty—something with real muscle.

That doesn't work too well for cartels, and it won't work at all for an international climate organization. As an aside, the Organization of Petroleum Exporting Countries (OPEC) has (fortunately) had a huge amount of trouble with discipline, and so Saudi Arabia has to do almost all the work. We need to do a lot better than OPEC, and we can.

I see three reasons for optimism beyond the collective benefits of climate stability and energy security. First, the fairness payments that I describe in the next chapter will reward poor countries enough that they will find it cheap to participate and possibly profitable. This eliminates many enforcement problems. Second, quite a few of the wealthier countries seem to be cooperatively inclined. Third, the world has an ultimate enforcement lever that can do the job.

Sea Turtles and Ultimate Enforcement

As Joseph E. Stiglitz explains in his recent book, *Making Globalization Work*, the law we can use as the ultimate enforcement of the international climate agreement has already been tested—on sea turtles. We will get to them shortly, but first recall that countries should rarely be subject to the ultimate enforcement. People pay their speeding tickets rather than chance wage garnishment or prison. That's what we want for the international climate agreement—something that's strong and consequently almost never used.

Sea turtle populations exist around the globe. Several of the eight species are endangered, and one species faces likely extinction. The United States passed a law forbidding importation of shrimp caught in nets without U.S.-style turtle-excluder devices, but the World Trade Organization (WTO) struck down this law as arbitrarily and unjustifiably discriminatory. The United States lost the case because it provided technical and financial assistance to countries in the Western Hemisphere, but not to four Asian countries. They filed the complaint.

The United States then spent five years working with the complaining nations, except for Malaysia, which refused to cooperate. Next, the United States reinstated its turtle-protection policy, and Malaysia again filed a complaint with the WTO stating that the United States was not entitled to impose any prohibition in the absence of an international agreement allowing it to do so. This time the WTO sided with the United States.

The WTO made this significant statement in its initial decision: "We have not decided that sovereign states should not act together bilaterally, plurilaterally or multilaterally, either within the WTO or in other international fora, to protect endangered species or to otherwise protect the environment. Clearly, they should and do." The WTO also noted that, under WTO rules, countries have the right to take trade action to protect the environment and exhaustible resources, and the WTO does not have to "allow" them this right. These decisions declare that countries can impose trade restrictions—even stopping unwanted imports entirely—to enforce their own global environmental policy. This means it's legal for countries to enforce a global pricing policy using trade sanctions.

The WTO's ruling allows for the ultimate enforcement of a global pricing policy. Trade sanctions have real teeth. They are just what we need.

The idea of enforcing a climate agreement with trade sanctions is not new. Harvard professors Richard N. Cooper and Jeffrey Frankel have written about the idea (Cooper in 2000 and Frankel in 2004). And Nobel Prize–winner Joseph E. Stiglitz explains the idea in his book *Making Globalization Work* (2007). Although enforcement based on trade policy would be equally useful with a system of national caps or a system of global carbon pricing, Cooper and Stiglitz recommend coupling it with global carbon pricing.

Frankel says, "Trade sanctions are perhaps the most powerful multilateral inducement that can be applied to shirkers, short of military force." Because they are such a strong measure, we should use trade sanctions cautiously. But they should be part of the system. We should use them to enforce fines, and they might induce holdouts to join the world climate agreement.

The oil-exporting countries, which often subsidize carbon use—for example, Iran and Venezuela subsidize gasoline—will be the toughest challenge. It is to their economic advantage to undermine the agreement and to thwart all efforts to conserve oil and gas. However, their gains from keeping the price low domestically are relatively small. Selling oil cheaply at home when they could sell it for a high price abroad cancels most of the advantage they get from subsidized consumption. Because of this, trade penalties might just do the trick and get them to cooperate—albeit grudgingly.*

⌒

Cooperation never comes easily, and when anyone can quit and still get 90 percent of the benefit, many will choose to honor their commitments in the breech. Two levels of enforcement are necessary to secure cooperation. The first is a simple and immediate penalty schedule for noncompliance. The second is a real threat that backs up the penalties. Trade sanctions can serve as the threat and are strong enough to rarely need using.

Primary enforcement of global carbon pricing is simply a modest set of rewards for countries that exceed the target and fines for those that fall short. Some countries will choose the reward, and others will choose to buy their way out of full compliance. This freedom to choose will make global carbon pricing more popular than it would be with heavy-handed enforcement.

chapter 27

International Fairness

It is reasonable that every one who asks justice should do justice.

—Thomas Jefferson

About one-quarter of the earth's population has no access to electricity or fossil fuel. It seems presumptuous to ask them to share, even in proportion to their small incomes, in solving problems that they have played no role in causing.

The Kyoto Protocol imposes no obligations on developing counties but allows them to sell carbon credits—for use in countries with caps—to developed countries. This addresses fairness, but it goes too far, as the U.S. Senate agreed when, in 1997, it passed a resolution by 95-to-0 opposing any treaty lacking obligations for developing countries. Also, as I explain in Chapter 23, selling credits for *not emitting* leads to gaming. This makes the Kyoto Protocol ineffective and expensive, as well as unacceptable to the United States.

Global carbon pricing is more fair to begin with than emission caps are, but it too will need adjustment. Poor countries should not have to tax their low rates of carbon emissions at the same rate as wealthy countries tax their high rates of emissions, unless we give the poor countries some financial assistance. Since the enforcement mechanism that I described in the previous chapter tends to make all countries set their price near the global carbon target level, assistance is in order in the form of fairness payments.

Fairness payments are calculated in two steps and are based on the (somewhat incorrect) assumption that all countries set their carbon price exactly at the target.[1] The first step is to use a sliding scale that transfers revenue from high-emission countries to low-emission ones. The second step prevents low-emission countries from collecting fairness payments if they don't comply with the policy or if they comply only minimally.

Although the purpose of these payments is simply to make the system fair, they confer additional benefits. First, the fairness payments make poor countries want to comply, which takes considerable pressure off the enforcement mechanism described in the previous chapter. Second, fairness payments encourage nonprice approaches to reducing emissions. I'll return to these effects after I explain how fairness payments work.

Step 1: The Sliding Scale

The sliding scale determines "fairness prices," which are used to calculate the fairness payments. It assigns higher fairness prices to countries that are richer and use more fossil fuel.

Fairness prices can be thought of as the carbon prices countries "should" adopt to be fair. It's more cost-effective, though, for all countries to use the same carbon price. That's why the sliding scale determines payments and does not push countries to actually implement the fairness prices. The only purpose of fairness prices is to calculate fairness payments.

The sliding scale assigns higher fairness prices to richer countries—more or less. It's not exact in this regard, because the sliding scale is based on carbon emissions rather than income. Richer countries emit more carbon, but not exactly in proportion to their income.

I have based the scale on emissions instead of income for two reasons. First, measuring income is difficult and contentious. Second, linking the payments to emissions provides a helpful incentive—which, as I said, I'll return to later. Emissions must be measured per person so that a large country is not unfairly assigned a high fairness price simply because it is large. Although other designs are possible, the simplest one—the one I describe—makes the fairness price proportional to a country's emissions per person.

Fairness Payments

It's easiest to explain fairness payments with an example. To keep it simple, I'll use approximate, round numbers. Suppose India emits 1 ton per person per

---

1. This is not done for simplicity, and it is not an approximation. It is done to prevent incorrect incentives. Fortunately, it also simplifies the design.

year, an average country emits 5 tons per person, and the United States emits 20 tons per person (see Table 1).

Also suppose the global target carbon price is $10 per ton. The average country is automatically assigned the target price as its fairness price. So that tells us that a country that emits 5 tons per person gets a fairness price of $10 per ton. Because fairness prices are proportional to emissions, the fairness prices for India and the United States are $2 and $40, respectively, as you can see by comparing the second and third columns of the table.

Table 1. *Calculating Annual Fairness Payments per Person**

	Emissions per Person	Fairness Price	Fairness Gap	Fairness Payments per Person
India	1 ton	$2	– $8	– $4
Average Country	5 tons	$10	$0	$0
United States	20 tons	$40	$30	$15

The fourth column shows the fairness gap, which I define as the fairness price minus the global target of $10 per ton. This assures that an average country has no fairness gap and makes no fairness payments.

Finally, we multiply the fairness gap by two scale factors. First, we multiply by the world's average emissions, which are 5 tons per person. Then we multiply by the reward/penalty rate, Z, from the previous chapter and again assume it is 10 percent.

The results are fairness payments per person, as shown in the final column of the table. The United States must pay $15 per person per year, and India will receive $4 per person per year. Over all countries, the fairness payments and receipts will exactly balance; the arithmetic ensures this.

Quite possibly this plan would be unpopular in the United States. But remember that a cap-and-trade system will almost surely evolve to allow the purchase of carbon permits or credits from abroad. Bills before Congress already specify this, and the pressure from business will mount as permits become more expensive. Purchasing permits or credits from abroad is like making fairness payments—disguised though they may be—to other countries.

Also, because China is already emitting at the world's average rate, we would not make payments to China, and they would not be obligated to make any fairness payments. Under the present system, China gets the largest share of revenues in the world from selling carbon credits. And here's one last thing to keep in mind: In 2008, the United States is spending about $1,500 per person on oil imports, a hundred times more than the fairness payments in the example.* I'd call this an unfairness payment.

Of course, with a higher global target price and a higher reward/penalty rate, Z, fairness payments would be higher. But the more important question may be, What good do these payments do? That raises a new issue. Can countries get these fairness payments even if they don't comply with the global-carbon-pricing policy? No, they can't, because of step two.

Step 2: No Free Rides

Once fairness payments are calculated, countries slated to receive payments are checked for compliance. If a country has achieved a national carbon price equal to the global-carbon-pricing target or greater, its fairness payment is not modified. But if it is undercomplying, the fairness payment is scaled back. And if a country's national carbon price is equal to its fairness price or less, the country receives no fairness payment. For example, India would receive no fairness payment if its national carbon price was $2 per ton or less. With a carbon price between $2 and the global target of $10, India would receive a prorated share of its fairness payment.

If any countries have their payments scaled back, the savings are passed through proportionally to countries with above-average emissions, so high-emission countries end up paying less.

What's in It for Us?

Altogether, half the world's emissions come from countries with below-average emissions—the countries that will receive fairness payments. These countries will likely contribute more than half the emissions in coming years and more than half the increased demand for oil. Fairness payments will help bring these countries into full compliance with the global-carbon-pricing policy.

India, for example, emits far less per person than the global average but has the world's fastest-growing population, which is expected to surpass China's by 2050. It is also growing economically at a tremendous rate. Bringing countries like India into full compliance will stop their subsidization of fossil fuel and cause them to increase its price instead. This will have a huge impact on the way their fossil-fuel use develops—much greater than the impact of paying them for carbon credits based on "additional" carbon-emitting projects they avoid as determined by the United Nations. And paying them to join the world effort instead of paying them for individual projects will likely prove far cheaper.

Also, participation in global carbon pricing will bring these countries into the global oil consumers' cartel, helping reduce world oil prices more than the U.N.'s clean development projects ever will. As developing nations join the effort to reduce oil use rather than subsidize it, the resulting reduction in the world price of oil may well cover the entire cost of fairness payments.

In short, fairness payments are probably the most cost-effective measure we can take to curb global warming. Developing countries have other, more pressing problems, and without fairness payments they will simply feel it is unfair for them to clean up the mess we got rich making—even though we intended no harm. Without today's low-emission countries, we cannot solve the global problems of climate change and energy security.

A New Reason to Reduce Emissions

In addition to increasing fairness, the sliding scale provides helpful incentives for reducing emissions. Previous proposals for a carbon pricing policy do not include a sliding scale. So if a national government reduces emissions by a billion tons of carbon per year, the country is still required to implement the same carbon price, and there is no change in fairness payment to reward it. Success is not rewarded. A sliding scale changes that.

With a sliding scale, a reduction in carbon emissions reduces a country's assigned fairness price. This saves money for countries making fairness payments and increases revenues for countries receiving fairness payments.

So fairness payments reward a country for reducing its per-person emissions. But what useful actions will this new incentive encourage that the carbon price incentive does not already encourage? A government might, for example, research the country's geology to find the best places for carbon sequestration. If this leads to more carbon sequestration, the country will be rewarded with a lower position on the sliding fairness scale. Or the government might replace traffic stoplights at some intersections with traffic circles. Insurance companies have shown that this saves lives, time, and gasoline. Or a country might conduct an advertising campaign to change attitudes and inform people how to save fuel more cheaply. Carbon pricing does not reward such actions, but all would reduce the country's carbon emissions, and the sliding fairness scale would reward them all.

How Fair Is Fair Enough?

Science cannot answer the question of exactly what is fair, so a political decision is necessary. The main thing is to reach an international agreement that includes at least all the major players. If this requires larger or smaller fairness payments than the ones I describe here, the sliding scale is easy to adjust.

~

A uniform carbon price across all countries is the most cost-effective way to curb emissions. But a carbon price is a burden. The burden on a poor country is proportionally less than the burden on a rich country, because the poor country generally has less carbon to tax. But even this system treats the very poorest countries unfairly.

A sliding scale can correct the problem without causing uneven carbon prices if it simply determines transfer payments. The easiest and most motivating way to use the sliding scale is to base it not on income, but on greenhouse gas emissions. This rewards every reduction of emissions, even those that a high carbon price does not directly encourage.

Most importantly, fairness payments can unite the world in a campaign for energy security and climate stability. Only with such a globally unified campaign will the world overcome its global challenges.

chapter 28

Carbon Pricing: What Counts?

Taxpayers are being asked to provide huge subsidies to oil companies to produce oil—it's like subsidizing a fish to swim.

—Massachusetts Congressman Edward J. Markey, 2006

NO ONE LIKES TO PAY FULL PRICE. And nations are no different when it comes to carbon. That's why we need a global policy. Because leaders don't like pricing carbon (or capping carbon), they will look for the easiest way that still counts to comply with global policy. So it matters what counts. If taxing vodka counts—alcohol does contain carbon—then countries will tax liquor more and gasoline a little less. That won't help the climate, because vodka, though a fuel of sorts, is not a fossil fuel, it's a biofuel.

For the most part, deciding what counts is about as simple as not counting the vodka tax. But a few subtler questions remain. In this chapter, I show how to resolve some of them.

Subsidies

Subsidizing oil is the reverse of taxing it. So in calculations to determine a nation's carbon price, fossil-fuel subsidies reduce the carbon price. A sensible carbon pricing policy deducts fossil-fuel subsidies from carbon revenues. Because of this, as I explain in Chapter 24, global carbon pricing takes a big step beyond the Kyoto Protocol, even if the global target price is zero.

Energy subsidies are common in developing countries, especially in those rich in fossil fuel. China spent over $20 billion in 2007 subsidizing gasoline. That means it has put a negative carbon price on gasoline. But even the United States still subsidizes fossil fuels, and pressure is mounting to extend even larger subsidies to fossil synfuels.

Counting subsidies properly—that is, negatively—shines a spotlight on them and removes most of the political incentive to provide them. For every dollar of subsidy, the government would need to collect a dollar of carbon tax. Why bother?

Cap-and-Trade Permits

Cap-and-trade permits sell at a price even when governments give them out for free. I discussed this in Chapter 23, but it is worth revisiting. What matters with permits is their market price, even when companies get them for free. If the owner of a coal plant needs 1,000 permits and gets 1,010 for free, it might seem as if the company would not have any incentive to use its coal more efficiently. But, in fact, it has exactly the same incentive as it would if it had to buy all its permits at the market price. Suppose the market price is $30 for a 1-ton permit. First, the company sells its 10 extra permits for $300. Then the plant manager thinks, "If I could save 100 tons of carbon, I would need 100 fewer permits and could sell them for $3,000."

So the motivation to save carbon depends on the market price of permits and nothing else. A $30 carbon tax provides the same incentive as requiring permits with a market value of $30. Coal plants save the same amount of carbon under either scheme, so both plans should count the same under a global-carbon-pricing policy.

Even if a company receives free permits, when it forfeits those permits to cover its carbon emissions, it is like paying a carbon tax. So administrators of a global-carbon-pricing system can check the market price of permits each month to estimate the value of permits forfeited. This value counts as carbon pricing revenue, just the same as carbon-tax revenue does.

Existing Carbon Pricing

What if a nation already has an oil tax or a cap-and-trade system in place when a carbon pricing system starts up? Is that counted? There is no need to punish good habits started in the past, so all carbon charges are counted, new or old.

Caps tend to punish the good and reward the bad. The better a country has done in the past, the more reasonable it seems to assign it a tighter cap—which is, in effect, a punishment. The same holds true when a program resets caps. If a country has "not been able to" meet its cap, that is an argument for a less aggressive cap in the future. A major problem with individually negotiated caps is that the system often punishes good behavior with a tighter cap and rewards bad behavior with a looser cap. We should not make this mistake with global carbon pricing.

Taxing Gasoline for Roads

What about a tax that a government places on gasoline for the purpose of paying for new roads? This is more difficult, because the tax may, in effect, be a tax on road usage and not on carbon. Sorting this out would be impossible in practice, so the architects of a global-carbon-pricing system should make an arbitrary decision and apply it uniformly across countries. Perhaps the simplest rule is to count all taxes on gasoline as carbon taxes.

Deception

Another problem is the possibility of deception. A country could tax gasoline at the pump but secretly subsidize oil refineries on a per-gallon basis. We must take this possibility seriously because, if it is easy to get away with, the international agreement would collapse. Fortunately, it is difficult to keep secret a multibillion-dollar subsidy.

Because the price of oil and the cost of refining oil are well known, it is not too hard to predict what gasoline should be selling for. If a country places a fifty-cents-per-gallon tax on gasoline but the price doesn't go up by fifty cents per gallon, it's pretty obvious something is fishy.

If a country is caught deliberately failing to report a carbon subsidy, the unreported amount should be tripled and subtracted from the country's carbon revenues. Because the enforcement mechanism that I describe in Chapter 25 is self-funding, any country that cheats harms other countries. Those countries not engaging in dishonest practices will have good reason to demand that deception be dealt with effectively.

~

As with any financial incentive or regulation, it's important to define carefully the rules for compliance and rewards. In this respect, global carbon pricing does not appear to present any unusually difficult hurdles.

Carbon subsidies are like negative carbon pricing and must be counted as such. Tax subsidies may be complex, but Congress keeps track of tax expenditures on a regular basis, so this is not a new problem. Permits must be valued at their market price. Existing carbon taxes should all count, and administrators of the program should detect and punish cheating.

None of this is particularly hard, and two aspects of the proposal give reason for optimism. First, countries with low emissions will likely gain a net benefit due to fairness payments. They will want to remain in good standing. Second, most countries stand to gain from the success of global carbon pricing.

chapter 29

A Consumers' Cartel

Foreign oil is costing us $500 billion a year. In 10 years, $5 trillion goes out of the country. It's nuts. It's the greatest transfer of wealth from one area to another in the history of the world.

—Oilman T. Boone Pickens, 2008

THE TWO GREAT ENERGY CHALLENGES—climate change and energy security—are converging in the political arena. Oil addiction is now seen as central to both challenges, and many other energy questions are now seen to overlap. But only one broad approach can meet both challenges at once. The world must reduce its use of fossil fuel. And by historical standards, it must do so with unprecedented speed.

Without deliberate action, change will come too slowly to meet the climate challenge and too dangerously to meet the challenge of energy security. Without deliberate action, we will unnecessarily transfer trillions of dollars to the exporting nations, which, by blind luck, own the majority of the world's oil and gas.

Both challenges are global, and to solve both requires an international organization. Such an organization is inevitably an oil consumers' cartel. It is also a gas and coal consumers' cartel.

Any organization of producers aimed at reducing supply is a producers' cartel. Any organization aimed at reducing demand is a consumers' cartel. A consumers' cartel brings precisely the changes we seek. By definition, it reduces consumption, as fixing the climate requires. And as the law of supply and demand predicts, it reduces the market price—the world price—of oil, gas, and even coal.[1] Reducing imports and lowering the price of oil lead to energy security.

To succeed we have no choice but to form a consumers' cartel. We can remain blind to this fact or we can embrace it. We can let the Organization of Petroleum Exporting Countries (OPEC) intimidate us into not saying "the dread words," as the *New York Times* called them in 1980. Or we can take full advantage of a cartel's benefits to unite the constituencies who most want to meet these two challenges—energy security and climate stability. So I say them again: Form a consumers' cartel. Learn to love those words and stop fearing OPEC. Protect our wealth and protect the climate.

Which Cartel Is Right?

OPEC has been a proud cartel from the start. Its purpose: to gouge the world, rich and poor alike. Moreover, at best its members make poor use of their spoils. *New York Times* columnist, and author of *Hot, Flat and Crowded*, Thomas L. Friedman has pithily described the result with what he calls the First Law of Petropolitics: "The price of oil and the pace of freedom always move in opposite directions in oil-rich ... states."

The purpose of an oil consumers' cartel would be to stop the gouging and save the climate. Between OPEC and a consumers' cartel, there is no question which one is right. Yet the policy of the United States for thirty years, ever since Henry Kissinger threw in the towel, has been "Don't bother OPEC, and no, no, no, we must never even mention the idea of having our own countercartel—a consumers' cartel."

Are we idiots?

Or is some powerful anticonsumer force actively influencing policy from behind the scenes—some force that would lose tens of billions of dollars a year if the price of oil returned to a conscionable level? I'm not one for conspiracy theories, but I have a hard time swallowing the idea that politicians and the public keep going so far wrong without a lot of "help."[2]

Is It OPEC, or Is It Nature?

As explained in Chapter 19, the Saudis, in 1979, cut back on their plans to increase oil production, and they have not increased their production since.

---

1. International coal shipments have been increasing rapidly and are now affecting the domestic price for coal.
2. It may be of interest that the National Petroleum Council, funded by the fossil-energy industry, is an advisory committee inside the U.S. government. It is part of the Department of Energy (DOE), brought in at the DOE's inception in 1977.

But perhaps the Saudis have less oil than they claim to have. They are secretive, so we don't know how much of their oil remains. It can only hurt them if we decide to prepare for a shortage of oil, so they would not want to warn us.

OPEC's power may be spent. Perhaps we are up against nature instead of OPEC; peak oil could be the real problem. Does that mean a consumers' cartel is a waste of time? Just the opposite. Cartels are typically used to manipulate a market in which the other side is free and competitive. It's harder to go up against an opposing cartel. The International Energy Agency now estimates that if supply and demand get much tighter, we are in for extravagant price increases. If nature prevents a supply increase, it works just as well to limit demand with a consumers' cartel.

It matters not at all whether OPEC constrains supply or nature constrains supply. Absent a consumers' cartel, the price goes up just as high, and the trillions still flow to OPEC and the oil companies. With a consumers' cartel, the price comes down.

Opportunity Knocks

In 2008, soaring oil prices again sparked outrage. But unlike twenty-five years ago, we have reason to hope for an international response. The Kyoto process and its successor have begun to organize the world as never before.

In fact, the world is practically begging for a consumers' cartel; people just haven't understood it. They simply know they want an agreement to use less oil. Because a consumers' cartel would lower prices and save countries such as the United States, China, India, Germany, France, and Japan tens of billions of dollars per year, a cartel motivates cooperation. It also provides the tangible benefits needed to quiet the acrimony of the international climate process.

It may seem a marvelous coincidence that one policy fix will take care of two energy challenges—energy security and climate stability. However, two extraordinary barriers stand in our way: the oil industry and environmentalists—strange bedfellows indeed. No, they have not joined forces or found a common cause. The oil industry blocks our path in order to protect its astronomical profits. Environmentalists, on the other hand—or at least many I've talked with—are simply confused about what a cartel does. A consumers' cartel would reduce oil prices, and lower oil prices will encourage oil consumption, and more consumption means more, not less, carbon emissions and global warming, or so many environmentalists seem to think.

Every step of this logic is correct in isolation, but here's the catch: A cartel does lower the oil price—the world price. But it's not the world price that drives consumption. Domestic oil prices drive consumption. A consumers' cartel must separate world and domestic prices, and we can do this by means of an oil tax or, better, an untax. So the real logic is this: A consumers' cartel would reduce *world* oil prices, but *domestic* oil prices will stay high to discourage oil consumption, and less consumption means fewer carbon emissions and less global warming. On thing is certain about a consumers' cartel.

If a consumers' cartel lowers the world price of oil, it will also reduce the world's oil consumption.

I can only guess at who has blotted this old idea out of the collective consciousness, but I suspect those with the most to lose from a tax on oil. I can't prove it, but it looks like the oil companies have brainwashed us all into believing that taxing oil wouldn't work or is impossible. OPEC at least is completely open about this and issues a new anti–gasoline tax report almost every year (see www.opec.org/library/Special Publications/Whogetswhat2008 .htm).

But a cartel would work, and that's why the oil companies and OPEC hate the idea. And it would be possible if we opened our eyes. And it would make us richer, not poorer.

Tax oil and give all the revenues to consumers. That's dirt cheap. Consumers consume less oil. Demand for oil falls, and so does its price. America pays exporting nations less. We are richer. Consumers pay extra for gasoline but get it all back—100 percent—with an untax. The United States, acting alone, could have a significant effect. But a consumers' cartel that organized most of the world could send the world price of oil tumbling—but only by keeping the domestic price of oil high.

It should surprise no one, least of all environmentalists, to find that since OPEC and the oil industry hate a consumers' cartel, environmentalists should love one. Opportunity is knocking, but it is up to all of us to understand what the oil companies hope will remain confusing.

How It Works

The international policies that I have described in Part 4 of this book are the policies of a consumers' cartel. But as I explain in Chapter 13, cartels fall apart when they cannot agree on caps and cannot enforce them. Historically, this has happened with production caps, because most cartels are producers' cartels. But the same holds for consumers' cartels.

Both history and economics tell us that cartel members will fight over caps and cheat on them. The Kyoto Protocol tried caps, and look what happened. Most of the cartel members fought for and gained exemptions from caps. Others, such as the United States, agreed on a cap and then quit the cartel. Some bargained for a loose cap as the price of joining. Others joined with good intentions and then "forgot" to comply. Not only was all this totally predictable, Kissinger's team foresaw it and discovered the remedy back in 1974. They called it a "floor price" for domestic oil.

The proposals I have put forth in Part 4 of this book apply the essence of Kissinger's remedy. I propose a uniform price on carbon for all members of the consumers' cartel. This avoids the disagreements caused by caps, which was Kissinger's main goal. Countries would be free to put a domestic floor price on oil as part of their compliance with the global carbon price. I have recommended this for the United States. For strategic reasons, having to do with OPEC's market power, minor modifications of this approach would likely be beneficial.

One hundred fifty national caps will lead to nothing but chaos and failure. One simple carbon pricing rule will do the job right. It would even work for a producers' cartel if it could monitor the price. But producers can easily keep the prices they sell at secret, so they must rely on quantity caps, which are easier to check up on but cause disputes. Consuming nations have an advantage because the prices at gas stations and for other retail transactions are public knowledge. So two things make price the instrument of choice for organizing a consumers' cartel. Price is both easy to agree on and easy to monitor.

Once a consumers' cartel is organized and functioning, how does it benefit the United States? As I explain in Chapter 19, the United States can reduce the world price of oil simply by implementing an untax on oil. If the United States alone puts and untax on oil that reduces the world price by $10 a barrel, this benefit accrues to every oil-importing nation. If two additional nations each used as much oil as the United States and put in the same effort—set the same tax rate, that would reduce the world price by $30, and this triple benefit would accrue to all oil-importing countries. A cartel multiplies our benefits without increasing our effort.

> ### Fighting OPEC's Market Power
>
> When OPEC raises its price, consumers eventually use less oil. This is one factor limiting OPEC's market power.
>
> With a domestic floor price for oil, OPEC can raise the world price to that level without consumers seeing any change in the price they pay. So OPEC knows consumers will not reduce consumption if OPEC sets its price to the floor price.
>
> Several factors make it difficult for OPEC to play this game, but a more dynamic floor price might curb OPEC's market power even more.
>
> This is an advanced design topic that I will discuss in more depth on stoft.com.

Organizing

Cartels fall apart. OPEC is the exception, but only because Saudi Arabia shoulders almost all the cost of accepting a low production cap. So how can a cartel bring about international cooperation?

The cartel idea consists of two parts, obligations and benefits. Members of a consumers' cartel are obligated to consume less than they would like, but they benefit from a lower world price. Cartels fall apart because quitting the cartel avoids the obligation but retains the benefit—in this case, a low world price—which is the same for members and nonmembers.

Of course, this cartel problem is exactly the issue faced by an international climate organization—which is a cartel in disguise. Members are obligated to cut consumption of fossil fuel. But countries would rather take a free ride, consuming as they like while enjoying the climate benefits secured by the cutbacks of others.

Recognizing we have an oil consumers' cartel adds no new obligations; it only reveals new benefits. Obligations tear cooperative organizations apart, while benefits glue them together. Counting the benefit of lower world oil prices helps strengthen the bonds of cooperation.

But does recognition of the cartel concept—including a recognition that it would reduce the world oil price—really change anything? Yes and no. No one benefits directly from understanding the concept. But currently, no one is counting the benefits of lower world oil prices, so those benefits are not acting as glue. In fact, well over half the available glue for an international climate organization is likely being wasted. The oil price effect is uncertain, but as I show in Chapter 13, it is large by most estimates. More importantly, lower world oil prices occur sooner than climate stabilization and are more tangible to most people. Especially in poor countries, the benefits of a better climate fifty or a hundred years hence pale in comparison with the benefits of cheaper heat and transportation in five or ten years.

Recognizing the benefits of a cartel also leads to a more effective international design as well as more-effective national policies. For example, the global-carbon-pricing policy that I recommend, unlike cap and trade, allows countries to emphasize oil over coal. Oil-importing countries will likely take advantage of this to tax oil carbon more heavily than coal carbon, increasing the impact of the cartel on oil prices. That, in turn, will make the program more popular in countries like China and the United States, because it will help unite the concerns of energy security and climate change.

The most difficult single step toward organizing the next stage of the Kyoto process will be to bring China and the United States into alignment. Together, they count for nearly half the fossil-fuel problem. For these two, climate change is more of a divisive issue than a unifying issue. China notes that we have caused far more of the problem, and we note that China is now ahead of us on emissions and that its rate of emissions is rising faster than ours.

Rather than blaming each other, why don't the two countries focus on an external problem instead? The problem of sending trillions of dollars to oil exporters is something we have in common. And it is a huge problem for both countries. The United States is running out of oil. We've gone from 40 percent dependent on foreign oil in 1974, when Nixon resolved to kick the habit by 1980, to about 60 percent dependent now. China is less dependent now but is expected to be 80 percent dependent by 2030. We have much in common and much to gain from cooperation.

The strategy of tamping down demand for oil by organizing a cartel will provide relatively quick mutual benefits. Reducing our own demand for oil will help us and help China. If China reduces its demand as well, that also helps both countries. Both nations will soon realize that the more they enlist the help of others, the more we all benefit.

Understanding that we can fight high gas prices and energy dependence with the same organization that helps us fight global warming changes the game completely. The new game brings immediate rewards that hit home. Since we are building a consumers' cartel anyway, why not do it right, making use of its full benefits and taking credit for them?

～

Like it or not, an effective international climate authority will act as an oil consumers' cartel. We should like it, and we should take advantage of it. It requires no additional effort. Acknowledging and advertising a cartel's advantages for oil-importing nations, such as the United States and China, would induce cooperation better than the threat of global climate change.

The price advantage, which people currently ignore, is no small matter. When a research team at the Massachusetts Institute of Technology studied carbon caps set for an 80 percent emissions reduction by 2050, the researchers estimated that the world price of oil would be cut by one-third in 2050. That could save the next generation hundreds of billions of dollars per year. Now the oil crisis looks worse, and the benefits of a consumers' cartel look even greater.

Fortunately, this shift to the cartel perspective is purely win-win. A cartel, though it lowers the world oil price, can only do so by reducing consumption and emissions. Unlike fossil-fuel supply policies, such as subsidizing synfuels, and drilling for oil, a cartel is 100 percent climate friendly.

Charge It to OPEC: By the Numbers*

Eventually, when we are all driving electric cars, reducing the price of oil will save us no money. But for the next few decades, it could save us a great deal. This calculation shows that, at least up to a 30 percent reduction in CO_2, a global climate policy is likely to be essentially free to the United States—paid for, in effect, by reduced payments to foreign oil companies.

This simplified example of a global climate-change program uses round numbers, which roughly reflect U.S. emissions and oil use in 2008 and 2009.

Assumptions:

- ► Preprogram CO_2 emissions are 6 billion tons per year.
- ► The untax rate is $50 per ton of CO_2.
- ► The program results in a 30 percent reduction in each type of fossil fuel.
- ► A 10 percent reduction in world oil use reduces the world price by 15 percent.
- ► The climate-change program covers three-quarters of the world's oil use.
- ► The world price of oil is $75 per barrel (bbl).
- ► The United States uses 20 million bbl/day and imports 12 million bbl/day.

Calculation of the social cost of the program to society:

1. Social cost = ½ × $50/ton × 30% × 6 billion tons = $45 billion per year.

Calculation of the savings from the reduced cost of oil imports:

2. World oil use decreases by 3/4 of 30% = 23%.
3. The world price of oil decreases by 1.5 × 23% = 34%.
4. Savings per barrel imported is 34% of $75 = $25.
5. A 30% reduction would cut oil use (and imports) by 6 million bbl/day.
6. Savings on the remaining 6 million bbl/day of imports would be:
       $25 × 6 million × 365 / 1000 = $55 billion per year.

Buying hybrid cars, paying extra for wind power in place of coal, and so on costs consumers an extra $45 billion per year. But consumers save $55 billion per year on imported oil. That savings is mailed to consumers as part of their untax refund checks in June of each year. As a result, the climate program consisting of a $50-per-ton untax on CO_2 is more than paid for by foreign oil companies.

Consumers will also pay $25 per barrel less for the 8 million barrels a day purchased from domestic oil producers. This provides an additional savings of $74 billion per year, which is generally not counted by economists because it is mostly a loss to American oil companies. But those not owning oil-company stock will enjoy their additional untax refunds.

Part 5

Wrap-Up

chapter 30

Finding the Path

More than any other time in history, mankind faces a crossroads. One path leads to despair and utter hopelessness. The other, to total extinction. Let us pray we have the wisdom to choose correctly.

—Woody Allen, commencement address, 1979

THE UNITED STATES, ONCE THE WORLD CHAMPION oil producer, is now in third place for production and twelfth place for reserves. As a nation, we are still the world champion oil consumers—by almost three times. In 2008, the price of oil is starting to shrink that gap but at a national cost of roughly half a trillion dollars a year. Our national energy policy costs a hundred times less and is doing very little.

We must choose: We can pay exorbitant tribute to the Saudis, the Russians, and the big American oil companies. Or, at long last, we can develop an effective energy policy. That shouldn't be a hard choice. If we decide to remain stuck in our fossil past, we will only end up paying more tribute. Instead, we should claim a new title, this time as champion of the next energy era—the low-carbon age. A good energy policy can do that and may save as much as it costs. At the same time, it limits the damage that the waning of the age of fossil fuel might cause.

But should policy discourage fossil energy, or should it promote carbon-free energy sources? Fortunately, the two tasks are flip sides of the same problem, and smart policy—carbon pricing—takes care of both sides at once. Carbon pricing raises the cost of everything fossil and raises the profitability of carbon-free energy at the same time.

Modern economics shows how the government can harness the market. A combination of government and market gives us the power to accomplish all that we need—but only if we combine the two properly. And therein lies the real problem—politics.

Only political action can balance government and the market for a quick and relatively cheap transition away from carbon. So good policy must concern itself as much with political barriers as with effective economics. Four political barriers play crucial roles:

- ► Ignorance of the need for change.
- ► The power of fossil profit centers.
- ► Ignorance of modern policy tools.
- ► Fear of the costs of change.

The first barrier, ignorance of the need, is crumbling. Most of the world understands the danger of climate change, and we all dread the price of oil.

The fossil profit centers—OPEC, Big Oil, and Big Coal—will be against us for decades to come, although the coal industry might switch sides if it embraces carbon-capture technology. But OPEC and Big Oil, with hundreds of billions of dollars at stake, will remain implacable foes of good policy. Worse, they are brilliant opponents and will continue to disguise their attacks as helpful policy suggestions. The only useful approach is to eliminate them from policy discussions, except as providers of data. This may sound harsh, but with so much at stake we cannot truly trust anything they say—though of course they will tell the truth when that suits their purposes. In any case, good policy does not require meddling in their industry; it requires only putting a price on their carbon. Consequently, they have little specialized knowledge to contribute.

This leaves two political barriers on which to focus: ignorance of policy tools and the fear of policy costs. These two barriers coincide because the guiding principle of modern policy design is cost minimization. And the best way to relieve the fear of cost is to minimize cost. In the end, the popular subterfuges for hiding cost will fail. Put simply, good policy means maximum bang for your buck. That's good economics, and it's even better politics.

That's the central point, but it's not the whole picture. Fairness matters, both internationally and nationally. And international cooperation is essential.

So far in this book, I've laid out policies and their rationale. This chapter lays out the step-by-step thinking behind the assembly of these policies in the hope this will provide a coherent framework in which to view them.

Walking the Path

To successfully navigate the path to climate stability and energy security, we must accomplish six major tasks along the way:

1. Develop an international carbon pricing policy without caps.
2. Develop a low-cost national policy.
3. Assign markets and government their proper roles.
4. Raise the price of carbon and send consumers a full refund.
5. Choose an untax over cap and trade.
6. Address fuel economy and energy research.

Let's check the reasons to follow this path, one step at a time.

Step 1: Develop an International Carbon Pricing Policy without Caps.

Between now and 2050, economic growth in developing countries—bringing with it increased oil use and emissions—could overwhelm any purely national policy we adopt. Without an international effort, it is unlikely that we can achieve energy security, and climate stabilization is impossible.

Reversing the United States' uncooperative international stance is a necessary step. However, it only returns us to the situation in 1997—the year President Bill Clinton signed the Kyoto Protocol.[1] Back then, the United States was cooperating, but the developing countries would make no commitments.

Any national energy plan that does not focus on engaging the international community is just tilting at windmills. Developing countries will not make commitments that prevent their citizens from reaching our standard of living. This rules out caps, at least caps that are tight enough to be effective. The only sound available option is an international carbon pricing policy. If a developing nation's carbon price is no higher than ours, it cannot block the country from attaining our standard of living.

The problem of international cooperation is primary, so we must begin with step 1: develop an international carbon pricing policy without caps.

Step 2: Develop a Low-Cost National Policy.

We have the technology to become completely energy independent and carbon free in the next twenty years. The trouble is cost—nothing else. But that cost is prohibitive. So we can go only partway. How far we go depends on how cost-effective our policies are. Even a lucky breakthrough would solve the problem by reducing costs. Cost is the limiting factor.

This means cost-effectiveness determines how much we accomplish. So the focus of a national policy must be to maximize our bang for the buck by minimizing the cost of saving oil and carbon. It doesn't matter what our spending limit is; the cheaper the cutbacks, the more we will cut back. So for step 2 we must develop the most cost-effective approach to national policy.

---

1. Because the Senate never ratified the protocol, we are not obligated by it.

Step 3: Assign Markets and Government Their Proper Roles.

To minimize cost, markets and government must both play the roles they play best. Markets minimize costs by using prices to organize millions of decision makers to make complex choices the government is poor at making. But only the government can see the costs that lie outside the market and send the right cost signals to the market. The government also needs to help consumers with long-range decisions and help investors with long-range risky research projects.

Step 3 is to choose the market-oriented path—to decide that, with the exception of a few advanced research and demonstration projects, the government will get out of the subsidy business. That means no more fossil subsidies or wind subsidies, no more ethanol subsidies, and no more renewable portfolio stealth subsidies. The next step, 4, provides the market-oriented replacement for all these and much more.

Step 4: Raise the Price of Carbon and Send Consumers a Full Refund.

For the bulk of energy policy, price can serve as the dividing line between roles. The government puts a price on carbon emissions, and the market finds the cost-effective way to save fossil fuel.

So price is key. That seems simple. We've all seen a million prices. And that's the problem. If the key to the energy crisis were brain surgery or quantum physics, we'd be home free. We'd call in the experts. But since the key is "price," the pundits all feel they're Einsteins.

Congress has now accepted the idea of cap and trade, and that's a carbon pricing policy—just what we need. But it is misunderstood. Prices play two roles: They pry money out of your wallet, and they influence your decision about what to buy. We all know this, but mostly we focus on the money pried loose—and that's the wrong focus. That's not why price is the right tool for energy policy.

When economists say to raise the price of carbon, it's not to pry money out of our wallets. In fact, they are quite worried about that part. But lobbyists love to see all that money flowing into government coffers where they hope to get their hands on it. Subsidies—that's what they're after, and the cap-and-trade bills are full of them.

But price is the key, not because it pries loose money but because it changes our decisions. It coordinates a million businesses and a hundred million consumers in a billion cost-saving activities. It beats government regulations by a mile when it comes to getting us to save oil and carbon in the cheapest way possible—because we know what works best for us.

So how can we can we obtain the benefits of better energy decisions based on prices that reflect the true costs of carbon and oil without letting the lobbyists pick our pockets? Send consumers a full refund. The key to this trick is not to send bigger refunds to those who use more oil—that would undo the good effects of carbon pricing. Instead, base refunds on the principle that we all have an equal right to climate stability and energy security. Send everyone exactly the same refund, in the mail, every June, just like in Alaska.

This solves the problem of energy policies that treat the poor unfairly. The poor spend a greater percent of their income on energy—heating their homes and driving to work. Even so, they spend fewer dollars than average on fossil fuel. So an equal dollar refund check that will send everyone the average amount paid will more than cover their costs from higher carbon prices. Unlike most energy policies, this one does not place the burden on the poor.

So the forth step is clear. Price carbon and return all the revenues collected on an equal-dollar-per-person basis.

Step 5: Choose an Untax over Cap and Trade.

The political trend is still toward cap and trade as the way to price carbon. Why fight it? This was a tough call. The advantages of an untax are listed in the next chapter, but two are decisive. Cap and trade inadvertently sends an antagonistic signal to developing countries. Cap and trade ignores the special needs of the oil market and energy security. Essentially, its just a developed-country environmental policy.

To succeed, the world must cooperate. That requires including the developed countries and those who are primarily concerned with energy security.

Fortunately, the untax is even a better environmental policy than cap and trade. And that's according to James E. Hansen. And he may be the world's most knowledgeable and ardent global-warming expert.

The untax is the silver bullet of energy policies. It can unite the energy security camp, the developing countries, and the environmental movement. And it requires no substantive compromise from the environmentalists. Step five may be the most important of all. Choose an untax.

Step 6: Address Fuel Economy and Energy Research.

Pricing carbon is the most important step, but price is not everything. Even with the right price, markets do not work perfectly. Consumers make systematic mistakes, particularly by undervaluing energy savings in the far future—over the life of a car or a house. Another market failure concerns advanced research. Some fundamental discoveries—for example, the steam engine—open many doors and lead to enormous social progress. But their inventors are under-rewarded by the market.

The sixth step is to address these market failures in a way that is market oriented although it cannot be as market-based as the untax. The race to fuel economy and a focus on advanced and risky research accomplish this.

Leaving the Past Behind

These six steps lead to policies that are not new to the policy world but are new to the political arena. It's time to move forward. Fortunately, current public discussion of cap and trade has opened the door to modern policies like the ones I propose in this book.

Of course, policy experts have reached no uniform consensus concerning the policies I have assembled. But, among experts, the main bit of controversy that remains is about the tilt toward carbon taxes and away from carbon caps. But as N. Gregory Mankiw, George W. Bush's former chief economist, wrote in September 2007, "Among policy wonks like me, there is a broad consensus [that] we need a global carbon tax." That is the conclusion of this book, and I agree that a consensus is forming.

At the other end of the political spectrum is Al Gore, who, in July 2008, said, "I have long supported a sharp reduction in payroll taxes with the difference made up in CO_2 taxes. This is the single most important policy change we can make." That is precisely Mankiw's position and only a little different from the untax I propose. In summer 2008, Hansen, who I just mentioned, advocated a carbon tax with a "100 percent dividend." That is identical to the untax.

As I recount in Chapter 16, many others from across the political spectrum back a refunded carbon tax. These include experts from the right-wing American Enterprise Institute; liberal economists Joseph E. Stiglitz and Paul Krugman; and William Nordhaus, the most prominent energy economist for the past three decades. Mankiw, Stiglitz, Nordhaus, and Cooper all agree on the necessity of implementing an international carbon pricing policy without caps.

In spite of the rapidly growing consensus among energy policy experts, a fear still remains, even among those who favor a carbon tax, that a tax is politically infeasible. That was, in fact, my greatest fear when I began to write this book in late 2006. But adding to the growing consensus among experts is an even stronger dynamic, which will overcome the negative politics of a carbon tax.

In Europe, as I explain in Chapter 15, electric utilities make billions in excess profits from cap-and-trade permits that governments grant freely. This is teaching not just Europe, but also the United States, a lesson. The body politic is realizing that granting free carbon permits to industry is a bribe of enormous proportions. The public and politicians of all stripes are rapidly adopting a new view—that governments should auction permits and return the revenues to the public.

For industry, auctioning permits completely changes the game. Industry loves cap and trade, not because it allows businesses some flexibility but because governments have handed them free permits of great value. The profitability of this scam has made putting up with the risks and hassles of permit markets worthwhile. But without the profits of free permits, cap and trade becomes just a risky, complicated form of carbon tax. Once the political process rejects free permits—and this will happen sooner or later—industry will switch sides and favor a carbon tax over a carbon cap.

Once industry starts backing policy wonks from across the spectrum, the political tide will turn in favor of an untax. So why wait?

~

Lacking an effective energy policy has sent trillions of dollars to OPEC and Big Oil over the last three and a half decades. And if we don't adopt an effective policy soon we could permanently damage the earth, fund a new generation of terrorists, and cost ourselves even more than in the past. Or perhaps we're feeling lucky.

Big Oil and OPEC think we should try our luck. Others think oil has peaked, so don't worry about the climate. Or perhaps we need crash programs, or to spend ourselves into the poorhouse. There are a million ideas out there, and some of them actually sound pretty convincing until you take a close look.

But after all the sound and fury, if you stand back and look at what we've learned it's not that mysterious, and it's not that scary. High energy prices do work. OPEC proved that beyond a doubt. Prices take a while, but not nearly as long as all the "quick fixes" that come and go.

So all we need is high prices that don't send money to OPEC and Big Oil or even to our government or special interests. We just need the prices to signal to business and consumers not to waste energy and not to pass up new alternative energy sources. So just tax carbon and oil and refund the money. And if world oil prices do shoot up again, give the oil tax a rest till they come back down.

With a respectably high tax rate this would solve most of our energy problems. Since all the money's refunded, a high rate is not painful. Of course a fuel-economy program based on competition, not government standards, would hasten the day we can plug in our cars. And spending more on energy research than on astronauts is an easy call.

So the best plan is to keep it simple, do what's obvious, and rely on the market to make the complex decisions.

chapter 31

The Complete Package

Let this be our national goal: At the end of this decade, in the year 1980, the United States will not be dependent on any other country for the energy we need.

—President Richard Nixon, January 1974

IN CARBONOMICS, I PRESENT a complete framework of national and international energy policies. The two sets of policies complement each other by design and address issues of both energy security and climate stability. In this chapter, I describe all the policies together, simplified for easy reference, along with a list of their advantages.

Previously, I have described national polices before explaining international policies as a way of starting on more familiar ground. Here I start with international policies, because they are essential and provide one reason for adopting an untax at the national level.

Think Globally First

If global energy policy is ineffective, we cannot make up for it with a good American policy, no matter how much we sacrifice.

I sometimes hear people say that if the United States does its part and adopts an emissions cap, China will follow. We tried that. At Kyoto, we agreed, subject to ratification, to cap our emissions near the 1990 level. China and all other developing countries said, "Fine, and you can pay us if you want us to help you out." In the issue of caps they have not budged since, and there is no indication they will if we adopt a cap now.

The Kyoto system is not working. No one is enforcing the caps (though a few are cooperating fully), and payments for helping out are subject to fraud and abuse, exactly as economics predicts. Expanding the Kyoto Protocol will exacerbate, not eliminate, existing problems. It was useful to get started, but it's time to learn from past mistakes.

While I cannot guarantee outcomes, this book presents the set of policies with the best chance of circumventing international roadblocks. The heart of the policy, global carbon pricing, is a standard idea and one advocated by top international experts concerned with Kyoto's failure. I have pushed it forward a bit to give it more flexibility, to make it more fair, and to broaden its incentives. The policy's purpose is to induce all nations to adopt a similar level of carbon pricing and to align the global average carbon price with a global target price, P. In a nutshell, here is the global carbon pricing plan:

The Global Target Carbon Price

▶ Each country gets a neutral score of zero if its average carbon price is the global target P. Higher or lower carbon prices generate positive or negative scores equal to the extra or missing carbon revenues.

Flexible Enforcement

▶ Each country receives a reward of Z times its score. This means that countries that underprice and thus have negative scores pay a fine.

▶ The reward rate, Z—say, 10 percent—is adjusted from year to year to a level that causes countries to price carbon at target P on average.

Fairness

▶ A country with average per-capita emissions is assigned a "fairness price" equal to target carbon price, P. Other countries are assigned higher or lower fairness prices in proportion to their emissions. These prices are used only to calculate fairness payments.

▶ Fairness payments are zero for a country with average per-capita emissions. Higher-emission countries pay lower-emission countries.

Ultimate Enforcement

▶ Once the process is in operation, countries that refuse to pay fines or join the system are punished via international trade sanctions.

This may appear complicated. But compared with repeated negotiation of caps for 180 countries, or global carbon permit and carbon credit markets, the plan is trivially simple. Fairness payment will be small compared with the international payments to developing countries for carbon credits that would be required under an equally effective Kyoto system.

This proposal does not specify a target for the global carbon price because such a price is political and largely subjective. Science gives us some guidance as to possible physical damage, and economics gives some guidance on the cost of the damage and the cost of different carbon price levels. We know even less about the benefits of energy security, though they could be enormous. Many people feel they know the scientific or moral answer to the question of how much global effort is necessary, but we are not even remotely close to a global consensus.

All of us can advocate for the level of effort, P, we feel is best, and we should. But the contribution I hope to make with this book is of another sort. As I see it, the main block to progress at the present is not the lack of consensus on a numerical goal for 2020 or 2050, but the lack of consensus on a global policy that could do the job when we do agree on a goal.

The system of global carbon pricing I propose has many advantages over the Kyoto approach. The approach I advocate focuses on inducing and strengthening the international cooperation that will generate success rather than presuming success and focusing on how rapidly we must achieve certain targets. The advantages of global carbon pricing are that it:

Does not require caps. Caps set relative to a country's 1990 or 2000 emissions would block developing countries from catching up. Naturally, they reject such caps. Avoiding caps avoids the politically impossible.

Requires all countries to commit. Committing to a carbon price is just as real as committing to a cap. Since a carbon price clearly does not prevent developing countries from catching up with developed countries, all nations can accept a price commitment.

Allows different methods of collecting revenue. Because compliance is based only on total carbon revenue collected, each country can choose its method of compliance: a cap-and-trade system, a carbon tax, or a carbon untax.

Allows countries to cushion oil price shocks. Countries are free to apply different carbon prices to different fossil fuels. This allows countries using a carbon tax to shift the tax away from oil and onto coal during a spike in the world oil price, cushioning their citizens against oil price shocks.

Facilitates a consumers' cartel. Countries suffering from oil addiction can set a relatively high carbon price on oil. Targeting oil enhances the effects of a consumers' cartel and reduces the demand for oil and the global price of oil. This encourages international cooperation, especially from the United States and China.

Provides flexible enforcement. The effectiveness of the policy depends on average compliance. If some countries wish to undercomply while, in effect, paying others to voluntarily overcomply, all are better off. Effectiveness on a global scale does not suffer.

Takes fairness into account. An equal carbon price is more fair than equal emission caps. But an adjustable sliding scale that requires high-emission countries to pay low-emission countries can likely achieve whatever fairness standard is necessary for international cooperation.

Equalizes carbon prices. Although flexible enforcement allows for variability in national carbon prices, the enforcement mechanism has a built-in bias toward uniform carbon prices, which are the most cost-effective. The sliding fairness scale adds to this effect.

Provides incentives for nonprice policies. The sliding fairness scale rewards countries that reduce their emissions per capita. So the policy rewards any government policy that reduces emissions. This encourages helpful policies that lie beyond the scope of carbon pricing.

Encourages cooperation instead of gaming. With national caps, most countries try, through stubbornness and backdoor dealing, to loosen their own caps and tighten the caps of others—that is, they try to game the capping system. With a global carbon price target, no country receives individual treatment, so all are in the same boat.

Eliminates carbon credits for unverifiable emission reductions. The Kyoto system pays less developed countries for not emitting what they would have emitted had there been no Kyoto agreement. Eventually, those amounts become unknowable, and already reports on the system's functioning are dismal.

All told, this simple international policy would be more stable, more acceptable, and more powerful than any variation on the Kyoto Protocol.

Act Nationally

Under the global pricing policy I have just described, nations can choose their own pricing policies. Which should the United States choose? I propose we adopt my Core National Energy Plan, which includes the following components:

A carbon untax

▶ The government taxes carbon and refunds all revenues to citizens on an equal-dollar-per-person basis.

A separate tax rate for oil carbon

▶ The untax rate on oil carbon should be low or zero when the world price is high and should increase when the world price falls.

A race to fuel economy

▶ The government scores cars on the number of gallons they will consume in a standard lifetime of, say, 120,000 miles. Carmakers receive rewards of something like $1.50 per gallon for gallons saved relative to the national average for a new car. Below-average cars fund the rewards of above-average cars, so there is no net cost.

Federal funding for research

▶ The government should increase funding for advanced research and cutting-edge demonstration projects by at least tenfold, to over $15 billion per year.

These policies all correct market failures—ways in which the private market fails to perform as it should. The policies fix the failures by addressing root causes rather than symptoms, to the extent possible.

For example, the market sets fossil-fuel prices too low, because it does not take into account the costs of climate change or of using the military to ensure oil security. The symptoms of this failure include things like too many private jets and Hummers. But instead of singling those out, the Core National Energy Plan addresses the root of the problem—low carbon prices—by raising the price of carbon.

The OPEC cartel causes a second market failure. The second part of the Core National Energy Plan addresses OPEC's price tampering by compensating for fluctuations in the world price of oil.

The third part of the plan addresses consumer myopia—the tendency of consumers to undervalue fuel cost savings that occur years in the future.

Finally, private markets under-reward fundamental research that has broad social benefits. The fourth part of the Core National Energy Plan addresses this last market failure.

The four parts of the Core National Energy Plan have many advantages over current policies and most of those now before Congress.

The untax is broad and simple. A carbon untax is applied upstream, at the sources of fossil fuels. Unlike Europe's cap-and-trade scheme, which covers only half of fossil-fuel consumption, the untax covers all carbon, with a hundred times fewer points to monitor and no complex permit trading markets.

The untax provides predictable incentives. Carbon-tax rates are more predictable than cap-and-trade permit prices, which are subject to market speculation and are thus highly volatile. So a carbon untax lowers investment risks and reduces the risk premium that investors will pass through to consumers. This also encourages investment.

Refunds make it cheap. Refunding all carbon-tax revenue makes carbon pricing cheap for consumers. The refunds allow carbon prices to be high enough to produce effective change.

Equal-per-person refunds are fair. The poor spend a greater percentage of their income (but fewer dollars) than the rich on fossil fuel. An equal-dollar-per-person refund helps the poor. It's also fair; it's like giving everyone an equal right to the atmosphere, the climate, or energy security.

Energy pricing harnesses everyone's ingenuity. Subsidies reflect special interests and the poor judgment of bureaucrats. But higher carbon prices encourage equally all methods of reducing carbon emissions. Businesses and consumers will join forces to seek out the cheapest, most effective methods.

A separate tax on oil carbon provides flexibility. A permit trading system equalizes the costs of coal and oil permits. But when oil prices spike, and oil carbon sells at ten times the price of coal carbon, oil should not be taxed. And when oil prices plummet again, oil carbon should be taxed more than coal carbon because of the security benefit of using less oil. A separate oil carbon tax rate does this, while carbon caps and permit trading prevent it.

A race cuts through red tape. Fuel-efficiency standards get tangled in bureaucracy because they are based on detailed estimates of the costs of future car innovations. No one really knows these costs, least of all the bureaucrats. A race requires no standards because car companies just try to beat their competition.

A race starts up quickly. Lawmakers must put standards in place years in advance to give companies time to adjust. There is no reason to delay the start of a race, and companies will implement many small changes that work quickly.

A race rewards all efficiency improvements. A company that beats a standard by a mile gains no reward. But in a race to fuel economy, every improvement earns a reward, not just improvements below some bureaucratically set efficiency level.

The race helps American car companies. The race to fuel economy rewards improvement rather than absolute efficiency levels. Improvement is easiest for the companies that start the farthest behind.

The plan funds research and development that the market can't handle. Certain types of research and development, especially advanced research, benefit an entire industry. When successful, such research is often under-rewarded by the market. The plan funds only these types of research, which the market dramatically underfunds.

The Choice Is Ours

For over thirty years, the United States has neglected energy policy, chosen policies that fail, and subsidized those who profit from our addiction. Inevitably, our dependence on foreign oil and our emissions of carbon dioxide have only increased—but with two notable exceptions.

Between 1978 and1985, the United States cut its use of oil more than 18 percent and its oil imports 46 percent. In spite of paying OPEC and Big Oil nearly $2 trillion extra to change our behavior, our national income grew 21 percent—faster than in recent years.

Together with the rest of the world we crushed OPEC, sharply cut fossil-fuel use, reduced carbon emissions, and kept the economy growing. So how did our president convince the country that implementing a policy as weak as the Kyoto Protocol—which would have reduced, not increased, our payments to foreign and domestic oil companies—would wreck our economy?

Once again we are cutting our use of oil, not because of our own policy but because we are being forced to pay extraordinary tribute to the oil barons of the world. In 1975, we ignored the potential of Kissinger's countercartel to reduce the world's dependence on oil and to avert the second (1979) oil crisis. This time, we have obstructed, rather than supported and improved, the Kyoto process, which has formed a weak consumers' cartel to limit the use of fossil fuels.

Partly as a result, oil prices have set new records, and in 2008 they are pumping half a trillion dollars a year out of the American economy. But again this foreign energy policy is working, and oil use is dropping. Money talks. Prices change minds and behavior. But there is no reason on Earth we should hand all that money to Big Oil companies, either foreign or domestic.

We have a choice. Since 1975, we have known enough to protect our-selves. But over the intervening years, we have learned much more about good policy, and the world has changed. Discoveries about the effect of fossil fuels on climate have united the world in a desire to reduce its dependence on fossil fuel. All we lack is effective leadership.

America could provide that leadership—if we choose wisely.

But effective world leadership requires wisdom, commitment, and the respect of the world community. We have the wisdom, and a growing number of our country's best minds are working to replace the misguided policies of the past. But they will not succeed without the help of an informed citizenry.

The United States must choose between solid economic policies and tough talk about what we will achieve in fifty years. What will work is exactly what OPEC rails against year after year. We must tax oil and tax carbon. This will slacken the world's thirst for fossil fuels. Then reduced demand will, as OPEC fears, slash the price we pay. And the revenues from our own oil and carbon taxes will remain at home.

So our choice comes down to this: continue to pay tribute to those lucky enough to own the world's fossil energy supplies, or charge ourselves the real costs of using fossil fuel—including the costs of climate change and energy security.

Why not charge ourselves so we can keep the revenues? If we lead the world in this direction, reduced demand for fuel will lower the world price for all. The excess profits we capture from OPEC and Big Oil will fund our transition to clean energy, a stable climate, and energy security. The choice is ours.

Endnotes

Each note is referenced by an asterisk in the main text.

Additional documentation is available at stoft.com under Carbonomics.

Chapter 1

Once upon a Time

Saving twenty Years' worth of U.S. Oil use. Although the price of oil changed more than the price of natural gas or coal during the OPEC crises, the United States began conserving all types of fossil energy. Twenty years' worth of U.S. oil use is a measure of all the energy saved, not just the energy saved by conserving oil. To verify that value, we can simply look at Figure 3 in Chapter 8—a graph from the Department of Energy. Note that, in 2000, consumers saved about 65 quadrillion Btu of energy, whereas they used only 40 quadrillion Btu of oil. Hence, in that year alone, U.S. consumers saved 1.6 times as much total energy as they used in the form of oil.

Saving Eight Years' worth of World Oil use. In the ten years from 1963 till the Arab oil embargo in 1973, the world's oil use more than doubled. The next year, the trend reversed direction, and during the next twenty-five years, through 2008, oil use increased only 50 percent. This appears to indicate that consumers saved much more than eight years' worth of oil production. But see the endnotes to Chapter 8 to read how I constructed a more cautious estimate. Of course, that caution means that my estimate may be too low.

Chapter 2

Wreck the Economy?

Figure 1. This figure is based on a study by MIT's Joint Program on the Science and Policy of Global Change, "Assessment of U.S. Cap-and-Trade Proposals" (report number 146), April 2007. The study found that, by 2050, gross domestic product would be 0.97 percent lower with a cap-and-trade policy than without, consumer welfare would be 1.79 percent lower, and "market consumption" would be 2.35 percent lower. To be cautious, I chose to show market consumption in the figure. (Consumer welfare, a slightly more meaningful number, includes the value of a small increase in leisure time.) The values are in 2005 dollars and are somewhat higher than typical median income values because they include the high incomes of the rich and the value of government services such as public schools.

Note that the climate policy reduces the growth rate of consumption by about 1/20 of 1 percent lower, not because a fixed climate program lowers the growth rate, but because the program becomes increasingly more effective and costly over the forty-year period.

The MIT report considers three cap-and-trade scenarios, and I have reported on the strictest of those. It is stricter than the following bills before Congress at the time of the report: Bingaman-Specter draft 2007, Feinstein 2006, Lieberman-McCain 2007, Udall-Petri 2006, Kerry-Snowe 2007. It is essentially identical to Sanders-Boxer 2007 and slightly weaker than Waxman 2007.

Economists' Statement on Climate Change. See endnotes for Chapter 14.

Chapter 3
Peak Oil or Liquid Coal?

Figure 3. I calculated Deffeyes's oil supply predictions from the logit equation on page 41 of *Beyond Oil*: $P = a\,Q\,(1 - Q/Q_T)$, where P is oil production, Q is cumulative output to date, a is a constant that Deffeyes estimates to be 0.059, and Q_T is the total amount of oil that the world will eventually produce. Deffeyes estimates this total to be 2.013 trillion barrels. He reports the two estimates on page 49 of his book.

Deffeyes's equation says nothing about when oil production will peak; it tells us only how high it will peak and how quickly it will decline. But on page 45 he tells us that production will peak in 2005 or the first few months of 2006. On page 3, he suggests that November 24, 2005—Thanksgiving—would be as good a guess as any. In spite of the fact that peak oil is his focus and that he reports the exact equation for the peak, Deffeyes never reports the level of peak production. Since he predicts it occurs when $Q = Q_T/2$, it is given by $a \times Q_T/4$, which comes to 81.35 million barrels a day. Two and a half years after the predicted peak, the world is producing 85.5 million barrels a day.

We cannot expect exact predictions, and production could well decline after 2008 in a fashion similar to what Deffeyes predicts. However, given Deffeyes's concern that we will need to "get acquainted with parsnips and rutabaga" during the five years immediately after his predicted oil production peak, he may be taking his own economics and oil-forecasting a bit too seriously.

Chapter 4
Is the Globe Warming?

Sinking islands. Chapter 5 of the 2007 report from the IPCC's Working Group 1 covers sea-level changes. It tells us that measurements show a "rate of sea level rise of 3.1 ± 0.7 millimeters per year over 1993 to 2003." So for the ten years before the evacuation of the Duke of York Islands that Gelbspan writes about, the rate of sea-level rise was about 1/8 inch per year. That's slightly faster than the average during the twentieth century, but nowhere near the 11.8 inches per year reported by Gelbspan. But as the IPCC tells us, the sea level can rise at different rates in different locations. Does that provide an explanation for the Duke of York Islands? Was the ocean rising ninety-six times faster than average there?

Generally, as the old saying goes, water seeks its own level. That's why lakes are flat. But due to winds, currents, and variations in air pressure, sea temperatures, and saltiness, the ocean level can deviate up or down from its average by a little. However, on page 409, Working Group I tells us that 0.15 meters, which is approximately 6 inches, is about as far as such deviations go. So perhaps, in one extraordinary year, half the change Gelbspan reports might occur. However, this would not repeat the following year. Water just doesn't run uphill that far. Besides, there is no reason global warming should cause a mound of ocean water around these islands.

In short, the IPCC reports indicate that if the Duke of York Islands need evacuating, it is not because of global warming. The only remaining possibility is that the islands are sinking. Several theories explain how this could happen, but the movement of tectonic plates seems most likely.

Chapter 5
Cheaper than Free

Hypercars and Race Cars. The quotations concerning Hypercars are from documents on the Web site of the Rocky Mountain Institute (RMI), www.rmi.org. "We are currently working with approximately 20 capable entities eager to bring Supercars …" is from a transcript of a 1994 interview with Lovins, publication T95-33. The next two quotes, "significant production volumes …" and "most, if not all, of the cars …" are from the same source.

The quote concerning "300–400-mpg four-seaters" is from "Reinventing the Wheels," by Amory B. Lovins and L. Hunter Lovins, as is the quote about "600 mpg." This article is also available on RMI's Web site.

The quote concerning the "1-plus-2-equals-10 equation" is again from publication T95-33.

The quote beginning "By Spring 1996 …" is from "Hypercars: The Next Industrial Revolution," by Amory B. Lovins, available as well on RMI's Web site.

Hydrogen Hypercars. Lovins's realization that "the Middle East would therefore become irrelevant and the price would crash" is from a 1995 interview with David Kupfer in *The Progressive.* Lovins forecast "the end … as we know them" of various industries in "Hypercars, Hydrogen, and Distributed Utilities," a 2000 slide presentation available on RMI's Web site.

In "Hypercars: Uncompromised Vehicles, Disruptive Technologies, and the Rapid Transition to Hydrogen," a 2000 slide show, Lovins predicted the wide availability of Hypercars by about 2006. He announced the development of his Revolution show car in 2001 at the Aspen Clean Energy Roundtable in a slide show titled "Critical Issues in Domestic Energy Vulnerability." Both slide show are available on RMI's Web site.

For a useful summary of progress on zero-emission vehicles, see "Status and Prospects for Zero Emissions Vehicle Technology," a report of the California Air Resources Board Independent Expert Panel 2007, available at www.arb.ca.gov.

Chapter 6

No Free Lunch

Hundred percent rational consumers? Human rationality is frequently misunderstood by both economists and their critics. The critics often mock the idea of rationality without understanding that economics often needs to assume that people are rational only on average. For example, we know that some people buy cars that are more fuel efficient than a rational choice would require, and other people buy cars that are less fuel efficient than a rational choice might dictate. If these two types of mistakes averaged out, there would be no argument for the government pushing consumers to choose more efficiency. Correcting the price of oil and carbon would be a sufficient policy.

Although consumers do make mistakes in both directions, economic theory does not tell us that these mistakes average out. But for complex reasons having to do with the social psychology of the profession, economists often make this assumption. In fact, this bias is so strong that an economist with a Ph.D. from a top-ten economics department told me once that human rationality is a logical necessity—if people weren't rational, that just wouldn't make sense. After an education in economics that never mentioned the scientific method, he simply could not imagine the possibility of human irrationality. But the rationality of humans is an empirical question.

In fact, a large body of empirical evidence collected by more scientifically inclined economists indicates that people have economically important biases that are not rational. Discounting of future energy costs appears to be among those.

Chapter 7

The Core Energy Plan

How much does it cost? The true cost of a carbon cap or a carbon tax is perhaps the most important economic concept in this book. It's the key to understanding why most economists believe energy policy can be fairly inexpensive. I explain the idea in more detail in Chapter 16, but here's a concise explanation

Taxing something causes people to use less of it. However, the tax itself is not a net cost to society, because the value of the tax revenue equals the cost of the tax—as long as the government does not waste the revenue. So the only real net cost comes when people pay to do things that they wouldn't ordinarily do—things that are cost-effective only because of the tax.

But the cost of doing things to avoid a tax is limited by the tax rate. People will not spend more than the tax to avoid the tax. As the tax rate on gasoline is increased from zero to $1 per gallon consumers will use less and less gasoline. But at each tax rate, they will spend only up to the tax rate to avoid using a gallon. So at a tax rate of $1 per gallon the average cost of avoidance is only about fifty cents. Hence the cost of a $1 gas tax is roughly fifty cents times the number of gallons avoided—not times the number of gallons used. But the government collects $1 times the number of gallons used. For this reason, until a tax has had a large impact, its real net cost is far less than the tax collected.

Chapter 8
Learning from OPEC

Figure 1. Because I was unable to find an econometric study of world oil use and OPEC's prices, the top curve in Figure 1 shows my own estimate of the path oil production would have taken had OPEC not raised prices in the late 1970s and early 1980s. But the longer it has been since OPEC's two oil crises, the harder it is to know what would have happened if those crises had not occurred.

Before the crises, every 1 percent increase in world GDP caused roughly a 1.5 percent increase in oil use. After the crises, the same change in GDP caused only about a 0.6 percent increase in oil use, and by 2005, the same change was causing only about a 0.4 percent increase. This decline may have been a response to the OPEC price shocks, or it might have occurred in any case. To make a conservative estimate, I assumed that the decline would have occurred in any case, in a smooth transition from 1.5 percent down to 0.6 percent during the OPEC crises. After the crises I assume the crisis had no effect on the link between percentage GDP changes and percentage changes in oil use.

Combining this smooth decline in the response of oil demand to changes in GDP with data on yearly GDP changes, I predicted the world's oil use in the absence of the OPEC crises. The oil prices are from British Petroleum statistics, and the figures for world oil supply are from the Department of Energy's Annual Energy Review 2007, Table 11.10. The savings through 2006 come to just over eight years' worth of world oil consumption at the 2006 rate of use.

Figure 2. Some of the supply response shown in Figure 2 might well have occurred without the OPEC price shock. So the supply response to OPEC may be an overestimate. However, it appears nearly impossible that it has been seriously underestimated by Figure 2. The values for non-OPEC oil supply come from the Department of Energy's Annual Energy Review 2007, Tables 11.5 and 11.6

Figure 3. The May 2001 report of Dick Cheney's National Energy Policy Development Group, National Energy Policy (available online at whitehouse.gov), explains the figure this way: "Since 1970, as the economy has shifted toward greater use of more efficient technologies, U.S. energy intensity has declined 30 percent." In fact, from 1970 through 1973, U.S. energy intensity increased. It only began its decline after OPEC raised oil prices. Notably absent from the explanation is any mention of OPEC, high oil prices, conservation, or the correct date of the change in trend.

Chapter 9
The World Oil Market

Kissinger's quote. Henry Kissinger, secretary of state under Richard Nixon, published and op-ed in the International Herald Tribune on January 18, 2007. He said in part: "American forces are indispensable. They are in Iraq not as a favor to its government or as a reward for its conduct. They are there as an expression of the American national interest to prevent the Iranian combination of imperialism and fundamentalist ideology from dominating a region on which the energy supplies of the industrial democracies depend."

Chapter 10
Corn Whiskey versus the Climate

Ethanol mileage. Currently, standard American cars go only two-thirds as far on a gallon of ethanol as on a gallon of gasoline. But a car that runs on only ethanol can be designed to get better mileage, because ethanol's higher octane allows it to be used with higher compression ratios. Of course, that still does not beat a diesel engine, which can have an even higher compression ratio. A gallon and a half of ethanol in a car designed for ethanol gets you no farther than a gallon of diesel in a car with a diesel engine.

Greenhouse gases from Ethanol. The referenced report from the *Proceedings of the National Academy of Sciences,* is "Environmental, economic, and energetic costs and benefits of biodiesel and ethanol biofuels," by Jason Hill, et al. Contributed by David Tilman, June 2, 2006, publshed July 25, 2006, in vol 103, no. 3, pp. 11206–11210.

Chapter 11
Synfuels Again?

Synfuel profits. On July 27, 2007, the online version of the *London Times* reported that "according to Shell's 2006 accounts, oil sands contributed $651 million in profits. … The synthetic crude made from dirty bitumen generates a post-tax profit of $21.75 per barrel, compared with Shell's average profit per barrel of crude of just $12.41."

Carbon emissions. Table 1 shows the relative carbon emissions from various fossil fuels. The carbon data for natural gas, crude oil, and U.S. coal are from an Environmental Protection

Agency document, "Inventory of U.S. Greenhouse Gas Emissions and Sinks: Fast Facts." The Department of Energy's Annual Energy Review 2007 provided data on the U.S. coal mix. The relative value of 1.0 for natural gas corresponds to 14.47 kilograms of carbon per million Btu of energy. Liquid coal is assumed to emit 1.8 times as much carbon as coal because of how much energy it takes to convert coal to a liquid.

Chapter 12
China, Coal, and Carbon Capture

FutureGen—a short history. On February 27, 2003, Secretary of Energy Spencer Abraham announced FutureGen. The text of his prepared statement, under the subhead "Carbon Sequestration and Hydrogen Power Plant," reads in part: "The Department of Energy, with private sector and international support, will embark upon a $1 billion initiative to design, build and operate the first coal-fired, emissions-free power plant. When operational, this plant—which we have named FutureGen—will be the world's cleanest, full-scale fossil fuel power plant. Using the latest technology, it will … provide a new source of clean-burning hydrogen."

Note the emphasis on hydrogen. Such a power plant is cleaner than a plant with a standard gas turbine only if the plant produces hydrogen. On May 24, 2006, President George W. Bush said, according to a press release available on whitehouse.gov: "We're developing clean coal technology. We're spending over $2 billion in a 10-year period … to determine whether or not we can have zero-emissions coal-fired power plants." Although he did not mention hydrogen, a zero-emission coal-fired power plant must make hydrogen. Even then, only about 90 percent of the carbon dioxide is captured—although that is still an excellent record.

Also note that Bush says it will cost over $2 billion. On January 31, 2008, the Associated Press reported that "[Deputy Secretary of Energy Clay] Sell said he and [Secretary of Energy Samuel W.] Bodman learned only last March [2007] that FutureGen's cost had escalated from an original $950 million to $1.8 billion. 'I knew (then) that we were into something that would not end well,' Sell told reporters in a conference call Wednesday." This seems peculiar, since ten months before Sell and Bodman "learned" of the $1.8 billion cost, and twenty months before the cancellation, Bush had said it would cost $2 billion.

More recent was the December 2008 news that Odessa, Texas, would not be the site of the FutureGen plant. On February 2, 2008, the *Wall Street Journal* reported that "Clay Sell, deputy energy secretary, said the easier, less-responsible path would have been to pretend everything was fine 'and then when the thing went south, I could have blamed the next administration.'"

When the Department of Energy scrapped plans for the FutureGen plant, it restructured the project and announced (in a press release available at www.doe.gov), "The restructured approach will focus on separating carbon dioxide (CO_2) for CCS, and does not include hydrogen production, which the concept announced in 2003 included."

Not only did the Department of Energy cancel the FutureGen plant it had been touting as its clean coal initiative for five years, but it also canceled funding for research on hydrogen production, which could make coal plants and hydrogen vehicles genuinely clean. Making natural gas from coal will never make coal as clean as natural gas, and it won't power hydrogen cars.

The *New York Times*, on January 31, 2008, explained that "about $50 million has been spent [on FutureGen], with about $40 million of that taxpayer money." That's $8 million per year. With the U.S. population at 300 million, that comes to less than three cents per person per year spent on solving the single biggest carbon dioxide emissions problem.

Chapter 13
Charge It to OPEC

Bush visits Saudi Arabia. George Bush senior's visit to Saudi Arabia is best described in the *Wall Street Journal*. On April 2, 1986 it reported "Oil prices sank below $10 a barrel to 12-year lows before rebounding sharply in response to a plan by Vice President George Bush to discuss oil-price stability with Saudi Arabian officials." On April 7, it reported his comment, "Hey, we must have a strong, viable domestic [oil] industry." The *Atlanta Journal-Constitution* discussed boiling him in oil on April 3rd.

How strong a consumer cartel? (The Economic Models.) Quite a few economic models predict that a consumers' cartel would be effective at reducing world oil use and lowering the price of oil. Here are a few examples.

The Department of Energy (DOE). As I discuss in the first two chapters of this book, the DOE's 1998 report on the impacts of the Kyoto Protocol estimated that it would reduce oil use, which would in turn lower the world price of oil: "Because of lower petroleum demand in the United States and in other developed countries that are committed to reducing emissions under the Kyoto Protocol, world oil prices are lower by between 4 and 16 percent in 2010, relative to the reference case price of $20.77 per barrel."

The 16 percent drop in the world price of oil corresponds to maximum compliance with the Kyoto Protocol and to about a 13 percent reduction in U.S. oil use. But I am interested in global effects, so I need a figure for the global change in use, not just the change in U.S. oil use. The 16 percent drop in the global price of oil corresponds to only a 3 percent drop in the global demand for oil. So the percentage price drop is five times larger than the percentage drop in global demand. That's an oil-change ratio of 1-to-5. (The oil use by "other developed countries" affects this calculation very little.)

Wharton Economic Forecasting Associates. Lawrence R. Klein, winner of the Nobel Memorial Prize in Economic Sciences, founded this economics forecasting organization, which was associated with the Wharton School of the University of Pennsylvania. Its prediction of the impact of the Kyoto Protocol included a prediction of the Protocol's impact on the world price of oil, which the DOE included in its report on Kyoto. The Wharton group predicted only a 13 percent drop in world oil prices under the same circumstances that the DOE used in its model. That's an oil-change ratio of 1-to-4.

The Electric Power Research Institute (EPRI). EPRI, the research arm of the electricity industry, predicted a 15 percent drop in the world price of oil under the same circumstances as those in the previous two models. So EPRI essentially agrees with the DOE's estimate of a 1-to-5 ratio. All three models predict that a small reduction in demand causes a relatively large reduction in price.

The Massachusetts Institute of Technology's Joint Program on the Science and Policy of Global Change. This group reported on a study of congressional cap-and-trade proposals in April 2007. It found that these proposals, when combined with weaker greenhouse gas initiatives in the rest of the world, would lower the world oil price by between 34 percent and 47 percent by 2050. Because the study does not report world oil use, I cannot compute the magnitude of the oil-change ratio these researchers used. However, the study assumes the world oil price would be $90 per barrel in 2050 without a climate policy and predicts that a strict policy would push the price down to $48 (in 2006 dollars).

Chapter 14
A Market-Based Carbon Tax?

Mankiw's op-ed. The op-ed by N. Gregory Mankiw that I quote in various chapters is titled "One Answer to Global Warming: A New Tax." It appeared in the *New York Times* on September 16, 2007.

Future caps. The Stavins report, which reviews the three optimistic evaluations of California's future cap, is "Too Good to Be True? An Examination of Three Economic Assessments of California Climate Change Policy," by Robert Stavins, Judson Jaffe, and Todd Schatzki, and is part of the John F. Kennedy School of Government working paper series, RWP07-16.

California's Global Warming Solutions Act, Assembly Bill 32, contains a loophole that allows some future governor to postpone its final goals indefinitely. The quote describing that loophole is from a press release from the Office of the Governor dated September 27, 2006, "Gov. Schwarzenegger Signs Landmark Legislation to Reduce Greenhouse Gas Emissions."

The first consensus. The Economists' Statement on Climate Change was originally drafted by Kenneth J. Arrow of Stanford University, Dale W. Jorgenson of Harvard, Paul Krugman of Princeton, William Nordhaus of Yale, and Robert M Solow of the Massachusetts Institute of Technology (MIT). Other signers include these Nobel laureates: Gerard Debreu and John C. Harsanyi of the University of California, Berkeley; Lawrence R. Klein of the University of Pennsylvania; Wassily Leontief of New York University; Franco Modigliani of MIT; Joseph E. Stiglitz of Columbia University; and James Tobin of Yale.

Chapter 15
Cap-and-Trade Politics

The cost of \$4,454 for a family of four. The \$4,454 direct cost of a carbon tax refers to the increased costs that a family of four would pay due to carbon taxes or carbon permits under cap and trade. It does not include refunds or reductions of other taxes. The value is based on the MIT study that I reference in the endnotes for Chapter 2.

The report considers three hypothetical cap-and-trade programs. The most strict of these has as a goal to reduce greenhouse gas emissions in a straight line starting in 2012 and ending in 2050 at a level 80 percent below the 1990 level. Because the study assumes permit banking is allowed, the actual reduction is greater in the early years and less in the later years. The cost of permits under a cap is equivalent to a tax on greenhouse gas emissions.

The appendix to that report lists the following values for this strictest scenario in 2015, three years after the start of the program. Permits cost \$53.17 per ton. National greenhouse gas emissions are 6,099 million metric tons per year. The population is 321 million. Because permit banking is permitted, the price of permits is high from the start.

Multiplying the price times emissions times 1.102 short tons per metric ton, then dividing by the population number, gives \$1,113.59 in revenues collected per person, or \$4,454 per family of four.

Chapter 16
An Untax on Carbon

The T word is anathema. This quote is from Kenneth P. Green, Steven F. Hayward, and Kevin A. Hassett, "Climate Change: Caps vs. Taxes," an American Enterprise Institute Environmental Policy Outlook, June 2007.

Table 1. I estimated the average hidden cost per person per year using a simple rule of thumb: The average cost of saving carbon is half the maximum sensible cost of saving carbon. With a carbon tax rate of, say, \$30 per ton, it does not make financial sense to spend more than that saving carbon—better to pay the tax. So, in this case, the average cost of saving carbon would be \$15 per ton. Table 1 also assumes carbon dioxide emissions of 22 tons per person, which is just a bit high in 2008. The formula for the per-person cost shown in each cell is then one-half times the percent reduction times 22 tons times the untax rate.

Chapter 17
Untaxing Questions

Huckabee's quote. Huckabee was discussing "this whole idea of carbon credits," and the quote is from the *Wall Street Journal*, August 18, 2007.

Do consumers care about the price. Nordhaus tells us that a 1 percent increase in price should lead to about a 0.24 percent decrease in the use of oil (see "The Global Rebound Effect" in Chapter 10). So a 400 percent increase in its price might cut oil use by roughly 62 percent—that's a 0.24 percent reduction compounded 400 times. (This is full effect would take years.) But a 5 percent reduction in income should reduce oil use by only about 5 percent. So the price effect is probably roughly ten times as strong as the income effect. And an untax should work about 90 percent as well as letting OPEC tax us. Given the amount of money saved, the untax is vastly superior.

Most estimates suggest that, per dollar collected, taxing coal does more to reduce carbon dioxide emissions than does taxing oil. So we can aim our own carbon untax at coal, a target the "OPEC tax" misses.

Chapter 18

Why Untaxing Is Fair

From theory to dollars. The values in Table 1 are from Summary Table 1 in a study by the Congressional Budget Office, "Who Gains and Who Pays under Carbon-Allowance Trading? The Distributional Effects of Alternative Policy Designs," June 2000.

Chapter 19

Taxing Oil—Double or Nothing

Bush visits Saudi Arabia. See endnotes to Chapter 13.

Figure 1. Figure 1 is adapted from Figure 26 in the Department of Energy's (DOE's) *International Energy Outlook 2004*. The figure also appears in the DOE's *Assumptions for the Annual Energy Outlook 2004,* which informs us that "OPEC oil production is assumed to increase throughout the reference case forecast, making OPEC the primary source for satisfying the worldwide increase in oil consumption." In other words, the DOE based its 2004 projections on an assumption that OPEC would nearly double production over the next twenty-five years, in spite of the fact that OPEC had reduced production in the two prior years and that OPEC's output was lower than it had been thirty years earlier.

Chapter 20

A Race to Fuel Economy

The NHTSA's view of consumer savings. The National Highway Traffic Safety Administration (NHTSA) gives the following explanation of how consumers and the NHTSA itself evaluate fuel savings:

The analysis does not rely on the value of fuel savings realized over the life of the vehicle. Our analysis considers the value of fuel savings realized in the first 4.5 years of the vehicle's life. The 4.5 year period is the average ownership period for new cars. We determined that the fuel savings during this period will be recognized and valued by light truck purchasers. Based on our analysis, which assumes that consumers value fuel savings over 4.5 years, there are net benefits for the average light truck purchasers. Thus, the average consumer will be willing to pay higher prices for improved fuel economy.

The quote is from the NHTSA's report "Average Fuel Economy Standards for Light Trucks Model Years 2008–2011," April 2006. This report is the NHTSA's final rule on light-truck standards.

Note that the NHTSA takes into account only the fuel-economy savings that occur during the first four and a half years of a vehicle's life. However, the NHTSA's recommendation for fuel-efficiency improvements rests on a calculation that these improvements will cost consumers less than what they save in fuel costs and that consumers will therefore be willing to pay for the cost of the improvements. In other words, the savings NHTSA requires pay for themselves even if you don't count any fuel savings after the car is four and a half years old.

Chapter 21
Crash Programs

Figure 1. Figure 1, which depicts federal research funding from 1961 through 2008, is from the National Science Foundation report "Federal R&D Funding by Budget Function: Fiscal Years 2006–08, Detailed Statistical Tables," August 2007.

Atom bombs and lunar landings. The best source of information I've found on Senator Lamar Alexander's Manhattan Project proposal is his speech to the Brookings Institution as part of its "Opportunity 08" series: "Energy Challenges for the Next President," Washington, D.C., May 12, 2008.

Green crash programs. The discussion of green crash programs is largely based on an article entitled "Fast, Clean, and Cheap: Cutting Global Warming's Gordian Knot," by Michael Shellenberger et al., *Harvard Law and Policy Review*, winter 2008. The information for the sidebar "Crash Program Polling" comes from a report by the Nathan Cummings Foundation, "New Poll Finds Hurdles, Opportunity on Global Warming," September 2007.

Chapter 22
The Great Cost Confusion

Op-ed by Monica Prasad. Prasad's op-ed "On Carbon, Tax and Don't Spend" appeared in the *New York Times* on March 25, 2008.

Chapter 23
Kyoto: What Went Wrong?

The You-First principle. The *International Herald Tribune* article is from June 14, 2008. The Chinese documents are at www.ccchina.gov.cn/en. The explanation of differentiated responsibility is from a talk by Su Wei, director-general of China's Office of National Leading Group on Climate Change, "Briefing on China's National Climate Change Programme," Bali, Indonesia, December 7, 2007.

Billions wasted. The paper I discuss in the sidebar is "A Realistic Policy on International Carbon Offsets," by Michael Wara and David G. Victor, Stanford University Program on Energy and Sustainable Development, April 2008.

The refrigerant HFC-23. For more information on international carbon credits and the greenhouse gas HFC-23, see "Truth about Kyoto: Huge Profits, Little Carbon Saved," by Nick Davies, the *Guardian*, June 2, 2007.

Chapter 24
Global Carbon Pricing

Who's for Global Carbon Pricing? Richard N. Cooper of Harvard wrote one of the earliest papers in favor of global carbon pricing, "International Approaches to Global Climate Change," *World Bank Research Observer* 15 (2): 145–72 (2000). But in this paper he mentions earlier work by William Nordhaus supporting a global carbon tax. A recent paper by Nordhaus, "To Tax or Not to Tax: Alternative Approaches to Slowing Global Warming," *Review of Environmental Economics and Policy* 1 (1): 26–44 (2007), continues his advocacy of internationally harmonized carbon taxes.

Joseph E. Stiglitz, in Chapter 6 of his 2007 book, Making Globalization Work, provides a description of global carbon pricing that is the closest of any I have found to the views I express in this book. N. Gregory Mankiw suggests a "global carbon tax" in his *New York Times* op-ed of September 16, 2007, "One Answer to Global Warming: A New Tax."

Chapter 25
Does the World Need a Cap?

Eighty percent by 2050. In trying to track down the myth that the Intergovernmental Panel on Climate Change (IPCC) set a goal of 80 percent carbon dioxide reduction by 2050, I asked author and environmentalist Bill McKibben for help. He told me that all kinds of scientists endorse the goal, but that the IPCC had not endorsed it "formally." However, the IPCC does not endorse anything informally. He also suggested www.1sky.org as a source of the myth.

The 1Sky Web site tells us, "We need to cut carbon at least 80% below 1990 levels by 2050" and footnotes that claim, saying:

Intergovernmental Panel on Climate Change, 2007. Fourth Assessment Report, Working Group III, "Mitigation of Climate Change." See Chapter 13, box 13.7, available at http://www.ipcc.ch/pdf/assessment-report/ar4/wg3/ar4-wg3-chapter13.pdf. Dangerous climate change is expected if concentrations of global warming pollution in the atmosphere exceed 450 parts per million and the global average temperature increases by more than 3.6 degrees F or 2 degrees C from pre-industrial levels. Avoiding these thresholds requires emissions reductions of 80-95% below 1990 levels by 2050 *in developed countries* [emphasis added].

This tells us that the 80-by-2050 goal is just for developed countries and that its aim is to hold the concentration of greenhouse gases down to a level of 450 parts per million (ppm). The IPCC report referenced by 1Sky confirms this. Section 13.3.3.3 of that report also tells us that "several studies have analyzed the regional emission allocations or requirements on emission reductions." But it does not say that the IPCC has agreed with the findings of these reports, nor even that the IPCC's Working Group 3 has agreed with the findings.

Working Group 3 has defined six categories of stabilization scenarios, and below Table 3.5 of its report, it says the bottom two categories have "very low targets." The 450 ppm target is the lowest in the category with the lowest targets. Only six out of 177 studies considered policies for this category. Environmental groups have selected this target based on predictions that it would hold global temperature increases to "2 degrees C from pre-industrial levels." Let us look more closely at this goal.

The 450 ppm stabilization scenario corresponds to an estimated long-run, or equilibrium, temperature increase of 2.1 degrees centigrade above preindustrial levels, according to Table 3.5 of the Working Group 3 report and Table TS.5 of the Working Group 1 technical summary. According to Figure 3.4 of the IPCC's synthesis report, the long run is a multicentury phenomenon, and according to Figure 10.4 of the Working Group 1 report, this means more than 300 years.

In 2000, the IPCC's "Special Report on Emissions Scenarios" (SRES) defined a number of standard emission scenarios. Of the six most prominent, the lowest was SRES scenario B1. Researchers estimated it would lead to a 1.8-degree centigrade increase in temperature from the time period 1980 to 1999 to the time period 2090 to 2099, according to Table 3.1 in the IPCC's synthesis report, which also notes that to convert to an increase over preindustrial levels, you would add 0.5 degrees centigrade. In other words, researchers predict that SRES scenario B1 would lead to a 2.3-degree increase over preindustrial temperature levels by about 2095.

The two scenarios—the 450 ppm equilibrium scenario and SRES scenario B1—require different policies. The 450 ppm scenario requires us to reduce emissions by 80 to 95 percent below the 1990 level by 2050. However, Figure 3.1 in the IPCC's synthesis report predicts we can achieve scenario B1 with a 40 percent increase in emissions between 2000 and 2050.

What explains this striking difference in policy requirements—an 80 percent decrease versus a 40 percent increase in emissions in 2050? First, scenario B1 results in a temperature increase of 2.3 degrees centigrade instead of 2.1 degrees. Second, if the 2095 concentration of greenhouse gases under scenario B1 is not reduced further, the temperature would continue to rise over the next few centuries, leading to a 3.3-degree increase (see Table 3.9 of the Working Group 3 report for the temperature predictions and Table 3.1 of the synthesis report, note c, for the greenhouse gas concentrations associated with scenario B1). Third, the policy requirement associated with the 450 ppm scenario is arbitrary. Many scenarios would result in a 450 ppm equilibrium in the long run, and some would call for less reduction in emissions by 2050 and more by, say, 2100.

In summary, although 80-by-2050 may be an appropriate target, it is the most extreme target in the IPCC report, researchers have barely studied it, and the IPCC gives no hint that it is the best target. And the case for this and other targets depends on unpredictable temperatures 300 to 500 years in the future.

Chapter 26

International Enforcement

Sea turtles and the World Trade Organization. The World Trade Organization (WTO) has documented this case on its Web site at www.wto.org/english/tratop_e/envir_e/edis08_e.htm.

Richard N. Cooper discusses the use of trade sanctions briefly in "Alternatives to Kyoto: The Case for a Carbon Tax," in *Architectures for Agreement: Addressing Global Climate Change in the Post-Kyoto World,* edited by Joseph E. Aldy and Robert N. Stavins (Cambridge University Press, 2007).

Jeffrey Frankel provides an extensive discussion of trade sanctions, including the WTO's sea-turtle decision, in "Climate and Trade: Links between the Kyoto Protocol and WTO," *Environment,* September 2005.

The most accessible treatment of this subject, and the one that inspired my discussion of sea turtles, is in Chapter 6 of *Making Globalization Work,* by Joseph E. Stiglitz (W. W. Norton, 2007)

.

Chapter 27

International Fairness

Carbon dioxide Emissions per person. Appendix A of the Department of Energy's *International Energy Outlook 2008* reports carbon dioxide emissions and population for 2005 and 2010. Linear interpolation gives the following values for carbon dioxide emissions per person in 2008: India emits 1.2 tons per person; China, 5.2 tons; and the United States, 21.6 tons. The world average is 4.9 tons of carbon dioxide emissions per person.

Cost of oil imports. Table 1.5 of the Department of Energy's *Monthly Energy Review September 2008* gives the merchandise trade value for petroleum and all fossil energy for the first seven months of 2008. Extrapolating to annual values and dividing by a 2008 U.S. population of 305 million gives a per-person net import cost of $1,380 for petroleum and $1,470 for all fossil energy, including petroleum. The difference is due to natural gas imports, which are growing and which OPEC and the world market will soon dominate, just as they dominate our oil imports today. Although oil prices were declining in the second half of 2008, the values that I report here will likely prove to be underestimates, because most oil is bought under long-term contract and not on the spot market. Note that $1,470 per person is $5,880 per family of four.

Chapter 29
A Consumers' Cartel

Charge It to OPEC by the Numbers. Step 1 uses the formula in the endnote for Table 1 in Chapter 16. Step 2 relies on two assumptions: that the global climate agreement covers nations consuming three-quarters of the world's oil and that the agreement affects all nations similarly. Step 3 uses the oil-change ratio of 1-to-1.5 that I discuss at the end of Chapter 13. Step 4 simply converts the percentage change in the world price of oil to a dollar change based on a world oil price of $75 a barrel. Step 5 relies on several assumptions: that U.S. oil use is 20 million barrels a day, that the agreement leads to a 30 percent reduction in all fossil-fuel use, and that every barrel saved reduces imports by one barrel. Step 6 simply computes the savings on 6 million barrels of imported oil a day and scales that savings up to an annual rate.

Of course, this calculation is necessarily approximate, but there is no reason to believe it overstates the expected savings. In fact, both the Department of Energy and the International Energy Agency seem to believe that, in a world with tight oil supplies, the oil-change ratio is higher than the value I've used, and so the corresponding savings would be greater.

Index